T0365977

Systems Approach to the Design of Commercial Aircraft

Systems Approach to the Design of Commercial Aircraft

Scott Jackson and Ricardo Moraes dos Santos

CRC Press is an imprint of the
Taylor & Francis Group, an **informa** business

First edition published 2021
by CRC Press
6000 Broken Sound Parkway NW, Suite 300, Boca Raton, FL 33487-2742

and by CRC Press
2 Park Square, Milton Park, Abingdon, Oxon, OX14 4RN

Library of Congress Cataloging-in-Publication Data

Names: Jackson, Scott, author. | Moraes dos Santos, Ricardo, author.
Title: Systems approach to the design of commercial aircraft / Scott
Jackson and Ricardo Moraes dos Santos.
Description: First edition. | Boca Raton, FL : CRC Press, 2020. | Includes
bibliographical references and index.
Identifiers: LCCN 2020010889 (print) | LCCN 2020010890 (ebook) | ISBN
9780367481742 (hardback) | ISBN 9781003053750 (ebook)
Subjects: LCSH: Aeronautics--Systems engineering. | Transport
planes--Design and construction.
Classification: LCC TL671.2 .J17 2020 (print) | LCC TL671.2 (ebook) | DDC
629.134/1011--dc23
LC record available at https://lccn.loc.gov/2020010889
LC ebook record available at https://lccn.loc.gov/2020010890

ISBN: 978-0-367-48174-2 (hbk)
ISBN: 978-0-367-51419-8 (pbk)
ISBN: 978-1-003-05375-0 (ebk)

Typeset in Times
by Cenveo® Publisher Services

Contents

Preface

Previous books, for example, Jackson (1997, 2015), focus on the use of the principles of systems engineering for the definition and development of commercial aircraft. This book takes a broader approach. It places the development of commercial aircraft in the broader scope of systems science and its derivative processes of systems thinking, systems approach, and finally systems engineering.

We also discuss the fields of complexity, systems architecting, and cognitive bias which are part of systems science. Terms from systems science include holism, emergence, hierarchy, and cohesion.

The good news is that aviation fatalities have fallen dramatically in recent years. The phenomena of complexity and cognitive bias have been shown to be factors in many of the remaining accidents. An understanding of these phenomena promises to bring the fatality rate even lower. The goal is that a deeper understanding of commercial aircraft in the context of systems science will contribute to that trend.

This book is an elaboration of material from workshops at aircraft companies worldwide. It does not contain any material that reflects internal processes in those companies; this book is intended to be generic.

REFERENCES

Jackson, Scott. 1997. *Systems Engineering for Commercial Aircraft.* Aldershot, UK: Ashgate Publishing Limited (in English and Chinese).

Jackson, Scott. 2015. *Systems Engineering for Commercial Aircraft: A Domain Specific Adaptation*, edited by Guy Loft. Second ed. Aldershot, UK: Ashgate Publishing Limited (in English and Chinese). Textbook.

OTHER BOOKS BY SCOTT JACKSON

Systems Engineering for Commercial Aircraft (Ashgate, 1997)

Architecting Resilient Systems: Accident Avoidance and Survival from Disruptions (Wiley, 2010)

Systems Engineering for Commercial Aircraft: A Domain-Specific Adaptation (Ashgate, 2015) (available in both English and Chinese)

Acknowledgments

We wish to acknowledge the assistance of Karin Mayer, MA and her knowledge of grammar, punctuation, spelling, and readability. In addition, Kenneth Cureton of the University of Southern California was helpful in the understanding of complexity and also cybersecurity.

Also, Denise Howard formerly of the University of Southern California was helpful in the understanding of systems thinking. Finally, co-author Ricardo Moraes dos Santos of Embraer can take credit for suggesting that our workshop material in Brazil could be converted into a book.

About the Authors

Scott Jackson is an INCOSE Fellow and has written three books on systems engineering: *Systems Engineering for Commercial Aircraft* (1997), *Architecting Resilient Systems* (2010), and *Systems Engineering for Commercial Aircraft: A Domain-Specific Adaptation* (2015) (available in both English and Chinese). His current focus is on consulting with both Embraer of Brazil and COMAC of China. Dr. Jackson was a lecturer at the University of Southern California for 10 years. He was a Senior Systems Engineer at Boeing for over 40 years.

His bachelor's degree in aeronautical engineering was from the University of Texas at Austin. His master's degree in fluid mechanics is from the University of California in Los Angeles. His PhD is from the University of South Australia with an emphasis on resilient systems.

Ricardo Moraes dos Santos is head of Systems Engineering at Embraer S.A. in Brazil. He is also INCOSE Brazil Chapter Director and INCOSE member. During his eight years of experience with commercial aviation, he has worked two years in the Chief Engineering Office with the Defense and Security Programs.

At Embraer, he has worked on the Ejets family of aircraft including the ERJ 170, ERJ 175, ERJ 190, ERJ 195, and the second generation of Ejets including the ERJ 175 E2, ERJ 190 E2, and ERJ 195 E2.

Between 2017 and 2019, he was involved with UAS operation studies, UAM (Urban Air Mobility) and Autonomous Flight (eVtol).

He is responsible for MISTI (MIT and Embraer) Internship Program Partnership in Systems Engineering Scope, STAMP (System-Theoretic Accident Model and Processes).

His current focus is on the system architecting process to provide new solutions to Embraer.

1 Vision

The idea of this book is to prepare the audience for the new perspective in a practical way, the authors put the vision about the entire experience of several years using systems approach to solve problems and create new opportunities for new systems in the commercial aviation scope.

This approach includes aspects that this book addresses, for example, safety, cybersecurity, architecture, and design decisions.

Another important point is the influence of human cognitive aspects on decisions. The authors explore cognitive bias as a cause of accidents. This book includes a discussion about important accidents and tries to connect those biases through new perspective about these accidents.

An aircraft, including the pilot, is a system, that is, a collection of parts that act together (the principle of *holism*) to achieve a stated purpose, that is, to deliver passengers and cargo to a destination safely and economically. The parts acting together achieve powered flight. The aircraft, including the pilot, is part of a larger system called the world aviation system. Even the aircraft with pilot and passenger loaded can't achieve its purpose without the support of enabling subsystems like the navigation system, the ground support system, and the loading and unloading support system.

To accomplish the purpose, the parts must individually be qualified by test, demonstration, analysis, or inspection to achieve its performance level and constraints, such as electromagnetic interference (EMI) and durability, and environments during all phases of the life cycle to include purchase, deployment, operation, support, and retirement. The qualification of the parts must also include the interaction among the parts (the principle of *interactions*).

The process for defining the aircraft is called the systems approach, which begins with identifying the stakeholder needs. Stakeholders for the aircraft include, at a minimum, the airline customers, the passengers, the aircraft manufacturer, and the regulatory agencies. The airline customers will define the minimum expected performance and cost goals. The regulatory agencies will establish the minimum acceptable safety levels. The aircraft manufacturer will demonstrate compliance with these safety levels (the process of *certification*).

After the stakeholder needs are identified, the aircraft itself is defined by two major processes systems, architecting and systems engineering. The purpose of systems architecting is to define the major parts of the aircraft both functionally and physically. For most modern aircraft, these parts consist of the wings, the fuselage, the empennage (the tail), and the propulsion. The purpose of systems engineering is to assure that the aircraft satisfies the stakeholder needs and that its component parts support this purpose.

Definition of the aircraft to achieve both performance and certification levels requires the coordinated and integrated efforts of all technical and managerial organizations. Technical efforts include the definition and qualification of all parts of the aircraft, including the interaction among the parts. The definition of each part includes the development of requirements for all life cycle phases. These requirements are the result of a flow down of requirements from the aircraft level (the principle of hierarchy). Aircraft-level requirements are the result of analysis and trade-offs of customer needs. This task is normally the product of the advanced design organization.

Lower-level requirements are the result of flow down and derivation of requirements at lower levels of the aircraft *hierarchy*. The requirements database that results from the parts requirements analysis is a major deliverable for the certification of the aircraft. The requirements database is normally a product of the systems engineering organization with inputs from design engineering and other technical organizations, such as the reliability organization, the production organization, and the maintenance organization. Design engineering also identifies the appropriate verification method for each requirement and assigns the methods to the appropriate organizations, such as the test organization. The safety organization plays a major role in establishing safety levels.

The net result of the requirements effort described above is quality requirements specifications without which a quality aircraft cannot be designed or built. The quality specification defines both the parts of the aircraft and the relationship of each part with all other parts, without which the aircraft as a whole entity cannot be defined. An inadequately defined single part or interface can result in an inadequately defined aircraft that may not meet its performance or certification goals.

Managerial efforts include program management to assure that all technical and managerial functions are performed together in an integrated way to achieve the aircraft purpose. Hence, integration is required for both the aircraft and the organization. Program management is also responsible for tracking technical and managerial risks and identifying and executing mitigation steps. Managerial efforts also include negotiations with airline customers to determine customer needs and to convert these needs into aircraft-level product requirements. These needs typically include range, speed, operating cost, durability, and other needs such as dispatch reliability. Managerial efforts also include management of the supply chain to include allocation of requirements to the suppliers and contractual supplier tasks (the supplier statement of work) and to validate supplier verification tasks. The contracts organization plays a major role in establishing contracts with suppliers that address both technical requirements and supplier tasks.

A feature of modern aircraft is *complexity*, which means that its parts interact in ways that are difficult to predict. Complexity is a major contributor to risk in aircraft development. A key player in managing this complexity is the systems architecting process, part of the systems approach.

Another consideration in modern aircraft is cybersecurity. The purpose of cybersecurity is to protect the aircraft from hostile electronic intrusions. Both systems architecting and systems engineering are required to meet this objective.

In summary, the vision is that the systems approach to the development of a commercial aircraft is an integrated effort from a product (the aircraft) point of view, the technical effort point of view, the managerial point of view, and the organizational point of view. The above are required to define a quality aircraft.

2 Systems

Meadows (2008, p. 11) has described systems in the most understandable way. She begins by describing a system in the simplest of terms: "A system isn't just any old collection of things. A 'system' is an interconnected set of elements that is coherently organized that achieves something." She clarifies that definition by pointing out that "a system must consist of three kinds of things: *elements, interconnections*, and a *function* or *purpose*." Other sources, for example, Dori et al. (2019) say that this definition only applies to human-made systems; however, Meadows asserts that the purpose of other types of systems can be deduced by observing their behavior.

This chapter explains the essential characteristics of a system, why an aircraft along with associated systems qualifies as a system, types of systems and how they are viewed (worldviews), how all of these concepts are part of systems theory, and the nature of derivative concepts including systems thinking, the systems approach, and systems engineering.

The first major systems thinker, according to Sillitto and Dori (2017), was Aristotle, who in the 4th-century BC observed that when many parts act together "the whole is greater than the sum of the parts." We will see in this chapter that an aircraft satisfies this definition.

More recently Ludwig von Bertalanffy (1901–1972) published his major work (Bertalanffy 1968) called General Systems Theory. Bertalanffy is regarded by many as the founder of modern systems theory. He concluded that a system is a set of "interacting units with relationships among them." The aircraft satisfies this definition also. For example, the avionics interact with many parts of the system including the human part.

TYPES OF SYSTEMS

As shown in Figure 2.1, the two primary types of systems are real and mental. Mental systems are sometimes called conceptual systems. Real systems can be either human-made or naturally occurring. An aircraft (without the pilot) is a human-made real system. The solar system is a naturally occurring real system. Works of art and literature are mental systems, not as the physical system that exists on a canvas or a sheet of paper, but rather the mental system in the mind of the artist. When a system consists of both human-made parts and humans, these systems are called sociotechnical systems. The flight manual is a mental system when only the information is considered.

Another subclass of systems is the abstracted system. The abstracted system is in fact a mental system that occurs in the context of a real system. In the commercial aircraft domain these types of systems are common. Every aircraft has, for example, various kinds of "trees" including specification trees, drawing trees, and reliability trees. These trees are, in fact, abstracted systems.

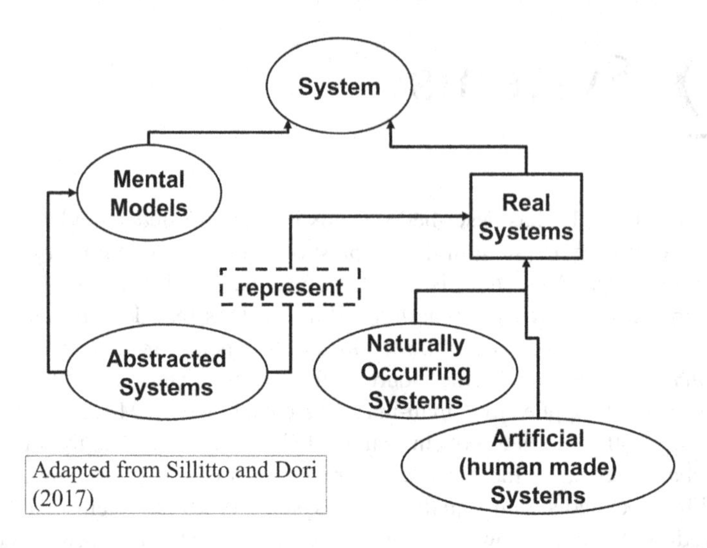

FIGURE 2.1 System types.

THE SYSTEMS VIEW

Other characteristics of a system that are often cited are as follows: first, the system should perform a function. This characteristic only applies to human-made systems. The function of an aircraft is pretty clear: it is to deliver passengers and cargo to a specified destination.

Cohesion

Another property of a system is cohesion. According to Hitchins (2009), cohesion simply means that all the parts work together to accomplish the stated function. If an engine fails to function, for example, the goal of achieving the aircraft function would be significantly impaired. Without cohesion an aircraft could not deliver its passengers and cargo.

Emergence

We saw earlier that Aristotle observed that the whole is greater than the sum of the parts. This property is called emergence and is one of the major properties of a system, especially an aircraft. The emergent property of an aircraft is *flight*.

Probably, the emergent property most familiar to the lay person is *wetness*. This is the property that results from combining hydrogen and oxygen atoms (the parts) to form the system called water.

Another possible emergent property that baffles even neuroscientists is consciousness. According to Graziano (2019, p. 34), one theory "sees consciousness as an emergent property of complex systems [the brain] and posits that the amount of consciousness in any system can be measured in units called phi."

The mechanisms that cause wetness and consciousness are beyond the scope of this book but suffice it to say that examples of emergence are all around us.

An example of emergence for aircraft is *flight*. Emergence simply says that you cannot achieve flight with only a part of the aircraft, the engine, for example, or a wing. You might be able to achieve a limited degree of flight with only one engine, but you cannot achieve the full potential of flight.

Another example of emergence for aircraft is safety. Although safety can be defined simply as the absence of failures, safety cannot be achieved without a specific arrangement of parts, as required by emergence. The same is true of cybersecurity.

Holism

Another characteristic of systems is *holism*, that is, the principle a system must be considered as a whole and that the whole is greater than the sum of its parts. Holism is particularly important for aircraft, for without holism, an aircraft would be just a collection of disconnected parts. For without connection, how would the aircraft be able to stay together and perform its function of transporting people and cargo safely to a destination?

Holism is also important for the achievement of safety, which was seen above to require all the parts to be connected, which is the essence of holism.

Hierarchy

Bertalanffy (1968) states that hierarchy is a property of all systems. However, hierarchy is often interpreted as a physical characteristic. Bertalanffy's meaning of hierarchy is not physical. He says that functional hierarchies are more appropriate. There are many nonhierarchical physical arrangements, for example, network systems and systems of systems are nonhierarchical systems. But even these systems can be depicted as functional hierarchies. The aircraft itself can be considered a system of systems when all the subsystems are acting together. This acting together is the key to continuous safe flight.

Returning to the subject of mental systems, many of the mental systems associated with an aircraft are hierarchical, for example, the specification tree, the reliability tree, the drawing tree, and so forth. One hierarchical structure that is universally recognized is the ATA 100 chapters (S-Tech 2016). In practice the ATA chapter system is most often used for maintenance accounting. However, it may be used for any process requiring a full accounting of aircraft parts. This full accounting brings us back to the concept of holism and the need to account for all parts of an aircraft.

According to Meadows (2008, pp. 82–85), "hierarchical systems evolve from the bottom up. The purpose of the upper layers is to serve the purposes of the lower layers."

Boundary

Figure 2.2 is a schematic of the solar system, which, of course, is a naturally occurring system that does not have a known function except to exist. Some writers, for example, Meadows (2008, pp. 11–12) maintain that natural systems have a purpose. However, it has multiple parts, the sun and the plants and the moons, and

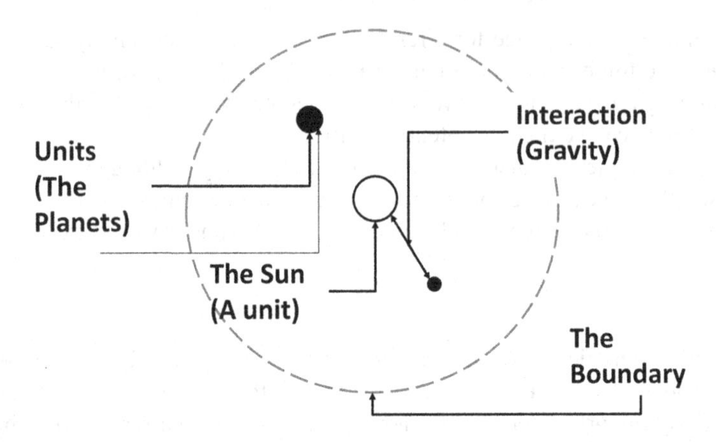

FIGURE 2.2 Solar system.

they interact through gravity. This chart also shows the theoretical boundary of this system when gravity can no longer hold it together. There is no universal agreement that all systems have boundaries.

In the aviation domain, there are many systems with ill-defined boundaries like the solar system. A collection of aircraft in a landing pattern is an example. It is in situations like this that create the risk for mid-air collisions and generate the need for continuous communication and visibility.

According to Meadows (2008, p. 97), "there are no separate systems. The world is a continuum. Where to draw around a system depends on the purpose of the discussion." This generalization pertains particularly to commercial aircraft and all its related systems.

TYPICAL SYSTEMS IN THE AVIATION DOMAIN

As discussed before, there are many other systems in the aviation domain than the aircraft itself. In accordance with the *interaction* characteristic described above, these systems interact within themselves and with the aircraft itself. Understanding all these systems is part of systems thinking.

The first system is the world aviation system. This system consists of all the aircraft in the world and the systems that control them. While individual aircraft may not interact with each other, they all interact with each other in the sense that they are all operating according to the same plan to direct and control them.

The second system is the air traffic control (ATC) system. This is the system that controls aircraft particularly when they are near the airport and are about to land or take-off. Hence, they comply with the interaction characteristic. This system has a direct interface with each aircraft that must be accounted for in the aircraft design. A primary purpose of the ATC is to assure that all the aircraft under its control do not collide with each other. Thus, it is seen that safety is the primary concern within all areas of the air control system.

Other systems include the enabling systems, such as the development system, the manufacturing system, the training system, and the maintenance system.

Another characteristic of a system is that it exists within an environment. Hence, the ATC, the airport, and the aerodynamic environment constitute the environment of the aircraft system. All of these aspects of the environment contribute to the safety challenge for the aircraft system.

The third system is the supply chain. This system does not directly interact with the aircraft during operation, but it is a critical system during development. Its interaction occurs during the development of subsystems and components. The supply chain is also part of the environment.

Then there is the aircraft itself. In this context, we should call it a socio-technical system since it includes the pilot, the crew, and all other personnel responsible for keeping it flying. In the broadest sense, it would also include maintenance personnel, baggage handlers, and ticketing personnel at the airport. Of course, as we have seen before, the parts of all these systems must interact with each other to preserve the system's function. This system, of course, has a purpose which we have seen is a basic characteristic of a human-made system.

This is the system which is the primary focus of concern for aircraft safety certification.

Figure 2.3 is a schematic of the entire aircraft system including the pilot, the ATC, and so forth.

The subsystems of the aircraft, such as the avionics, also qualify as systems since they perform a function and consist of many interacting parts. For the most parts, these systems cannot function independently without interacting with the rest of the aircraft. Figure 2.3 also shows these subsystems. In a regulatory context these subsystems are called systems, which we have seen is correct.

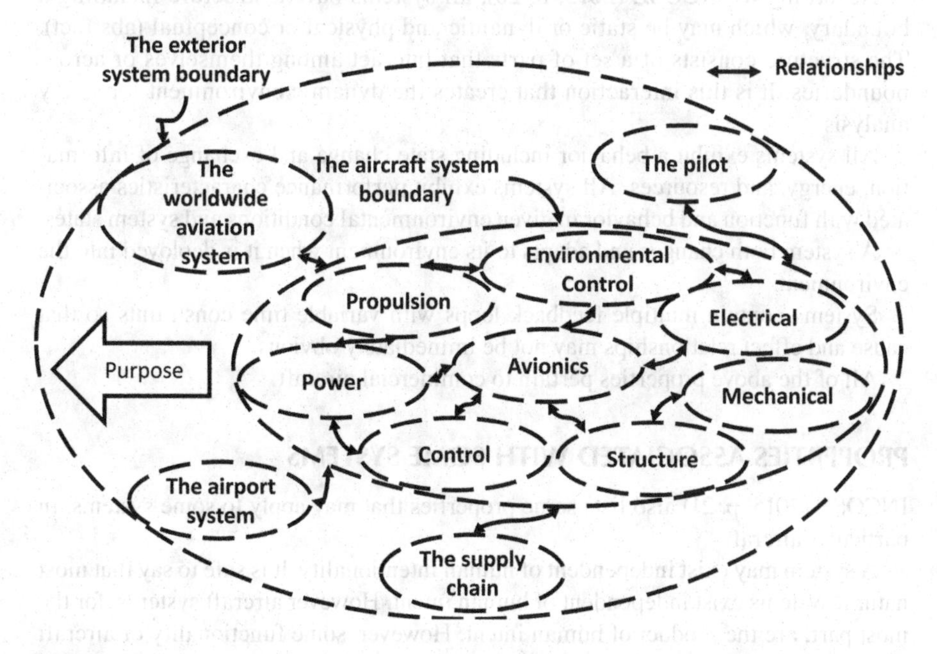

FIGURE 2.3 Aircraft as a system.

Finally, the components of the aircraft to the extent that they consist of multiple parts also qualify as systems.

Hence, it is seen that a commercial aircraft system itself consists of multiple layers of systems, each qualifying as a system in its own right. The importance of this fact is that the safety and operation of the individual aircraft are dependent on the total interaction and cohesion of the associated systems.

OTHER PROPERTIES OF SYSTEMS

Other properties of systems that are often cited include the following: understanding all these systems is part of systems thinking and is essential to the definition and development of successful and safe aircraft.

According to INCOSE (2015, p. 20), all systems especially human-made systems have a life cycle. According to INCOSE (2015, p. 29), depending on the domain, these life cycles typically begin with the determination of the user needs, continue through concept development (via systems architecting), development, production, utilization, and retirement. It is a continuous concern that each of these phases consists of multiple design decisions. The accuracy of these decisions is driven by the possibility of human error. One of the main sources of human error is cognitive bias discussed in Chapter 8.

According to Hitchins (2003), systems perform a function. For commercial aircraft this function is to deliver passengers and cargo to prescribed destinations. Hitchins' functions include "operate," "maintain viability," "manage resources," and "observe, orient, decide, and act."

According to INCOSE (2015, p. 20), all systems have a structure including a boundary, which may be static or dynamic and physical or conceptual (abstract). The structure consists of a set of parts that interact among themselves or across boundaries. It is this interaction that creates the dynamic environment for safety analysis.

All systems exhibit a behavior including state change and exchange of information, energy, and resources. All systems exhibit performance characteristics associated with function and behavior in given environmental conditions and system states.

A system both changes and adapts to its environment when it is deployed into the environment.

Systems contain multiple feedback loops with variable time constraints so that cause and effect relationships may not be immediately obvious.

All of the above properties pertain to commercial aircraft.

PROPERTIES ASSOCIATED WITH SOME SYSTEMS

INCOSE (2015, p. 21) also lists some properties that may apply to some systems, in particular aircraft.

A system may exist independent of human intentionality. It is safe to say that most natural systems exist independent of human intent. However aircraft systems, for the most part, are the product of human intent. However, some functionality of aircraft systems may exist independent of human intent.

A system may exist within a wider system. This is particularly true, for example, when an aircraft exists within the worldwide aviation system. This is another example of systems thinking.

A system may offer "affordances" features that provide additional advantages, intended or not. For example, most aircraft can float. This feature allowed US Airways to ditch in the Hudson river and save 155 lives, according to Pariès (2011, pp. 9–27).

A system may be clear and distinct from its context system. Aircrafts are a good example. That is to say, the ATC system will exist even if an individual aircraft system did not exist.

Some systems may be closely coupled with its context system. Once again, aircrafts are closely coupled with the ATC system.

Some systems may consist of a set of constituent systems. In the military domain, aircrafts are sometimes required to fly in formations. In the commercial domain this is rarely required.

Systems may require multiple disciplines for development and operation. This is certainly true for commercial aircraft. In addition to the traditional engineering disciplines, other disciplines, such as human factors and psychology are required. Psychology is an important consideration in avoiding cognitive bias that has been shown to be a source of flawed decisions in design and operation.

THE HIERARCHY OF SYSTEMS THEORY

As shown in Figure 2.4, systems theory can be seen to decompose into many areas of scientific study. As will be seen below, systems engineering is one of those areas. We will discuss systems thinking, the systems approach, and all the constituent parts of the systems approach.

Some of the areas of study discussed below are not often used in the commercial aircraft domain. However, in the interest of completeness, we will discuss them here.

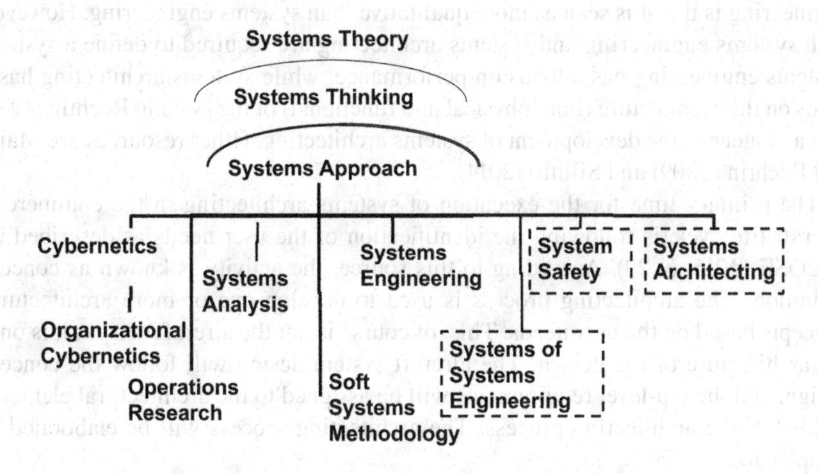

FIGURE 2.4 Systems theory.

These areas include cybernetics, organizational cybernetics, systems analysis, systems dynamics, and soft systems methodology. This is not to say that these areas are not useful; it is only to say that they are not often practiced in the commercial aircraft domain today.

SYSTEMS ENGINEERING

Of all the subordinate disciplines of systems science, systems engineering has received the most attention in the world of physical systems, their definition, and their operation. ISO/IEC (2008) is probably the preeminent standard for the conduct of systems engineering. Other volumes, such as Jackson (2015), have treated systems engineering in the aviation domain. However, these volumes do not discuss other aspects of the systems approach deeply, such as systems architecting and systems-of-systems engineering.

Systems in the Broader View

In the current SE paradigm, systems are robust, dependable, mainly technological, and deterministic. More recent researchers have begun to see systems engineering in a much broader view. Among these is Sillitto et al. (2019). According to this view, in the future systems will be evolutionary, resilient, and will encompass products, services, and enterprises. They will integrate technological, social, and environmental elements.

Future systems will involve environmental aspects, and certainly social aspects as well as engineering and technology.

SYSTEMS ARCHITECTING

Another branch of the systems approach that has come to the fore in recent years is systems architecting. The reason that this field is seen as separate from systems engineering is that it is seen as more qualitative than systems engineering. However, both systems engineering and systems architecting are required to define a system. Systems engineering has a focus on performance, while systems architecting has a focus on the architecture (both physical and functional) of the system. Rechtin (1991) was a pioneer in the development of systems architecting. Other resources are Maier and Rechtin (2009) and Sillitto (2014).

The primary time for the execution of systems architecting in the commercial aircraft life cycle is following the identification of the user needs as described by INCOSE (2015, p. 29). According to this source, the activity is known as concept definition. The architecting process is used to develop one or more architectural concepts based on the user needs. This, of course is not the aircraft design; it is only the architecture of the design. The aircraft system design will follow the concept design, and the top-level requirements will be assigned to the architectural elements defined in the architecting process. The architecting process will be elaborated in Chapter 10.

Another benefit of systems architecting is that it is useful in managing the complexity of the system. This aspect will be discussed more thoroughly in Chapter 5.

Systems of Systems (SoS) Engineering

This is another branch of the systems approach that considers multiple systems called component systems (CS) working together to achieve a common goal. The normal definition considers systems that when working independently are systems but having been brought together are now working together to achieve a common goal. The commercial aviation domain is replete with systems of systems. The aircraft working systems together with the ATC system is a system of systems. Jamshidi (2009) has the most complete examination of systems of systems.

REFERENCES

Bertalanffy, Luwig von. 1968. *General Systems Theory: Foundation, Development, Applications*. Revised ed. New York: George Baziller.

Dori, Dov, Hillary G. Sillitto, Regina Griego, Dorothy McKinney, Eileen Arnold, Patrick Godfrey, James Martin, Scott Jackson, and Daniel Krob. 2019. "System Definition, System Worldviews, and Systemness Characteristics." *IEEE Systems Journal*. Preprint (accepted for publication).

Graziano, Michael. 2019. "What is Conciousness?" *New Scientist* 243(3248):34–37.

Hitchins, Derek. 2003. *Advanced Systems Thinking, Engineering, and Management*. Norwood, MA: Archtech House.

Hitchins, Derek. 2009. "What Are the General Principles Applicable to Systems?" *Insight* 12(4):59–63.

INCOSE. 2015. "Generic Life Style Stages." *Systems Engineering Handbook*, edited by SE Handbook Working Group, 29. Seattle, CA: International Council on Systems Engineering.

ISO/IEC. 2008. *Systems and Software Engineering—System Life Cycle Processes*. Geneva, Switzerland: International Organisation for Standardisation/International Electrotechnical Commissions.

Jackson, Scott. 2015. *Systems Engineering for Commercial Aircraft: A Domain Specific Adaptation*, edited by Guy Loft. Second ed. Aldershot, UK: Ashgate Publishing Limited (in English and Chinese). Textbook.

Jamshidi, Mo. 2009. "Systems of Systems Engineering: Innovations for the 21st Century." In *System of Systems Engineering: Innovation for the 21st Century*, edited by Mo Jamshidi. Hoboken, NJ: John Wiley & Sons.

Maier, Mark W., and Eberhardt Rechtin. 2009. *The Art of Systems Architecting*. Third ed. Boca Raton, FL: CRC Press (Original edition, 1991).

Meadows, Donnella H. 2008. *Thinking in Systems*. White River Junction, VT: Chelsea Green Publishing.

Pariès, Jean. 2011. "Lessons from the Hudson." In *Resilience Engineering in Practice: A Guidebook*, edited by Erik Hollnagel, Jean Pariès, David D. Woods, and John Wreathhall, 9–27. Farnham, Surrey: Ashgate Publishing Limited.

Rechtin, Eberhardt. 1991. *Systems Architecting: Creating and Building Complex Systems*. Englewood Cliffs, NJ: CRC Press.

S-Tech. 2016. *ATA Chapters*. Limerick, PA: S-Tech Enterprises LLC.

Sillitto, Hillary G. 2014. *Architecting Systems: Concepts, Principles, and Practice*, edited by Harold "Bud" Larson, Jon P. Wade, and Wolfgang Hofkirchner, Vol. 6, *Systems*. London: College Publications.

Sillitto, Hillary G., and Dov Dori. 2017. "Defining 'System': A Comprehensive Approach." IS 2017, Adelaide, Australia.

Sillitto, Hillary G., James Martin, Regina Griego, Dorothy McKinney, Dov Dori, Scott Jackson, Eileen Arnold, Patrick Godfrey, and Daniel Krob. 2019. "System and SE Definitions." International Council on Systems Engineering, accessed 8 August. https://www.definitions.sillittoenterprises.com/.

3 Systems Theory

Systems pioneer, for example, Checkland (1999, p. 5) has pointed out that the reason the study of systems has had so much difficulty theory being understood is that it is not about objects themselves, but rather about other objects, their properties, their organization, and their function. Systems thinking is a collection of thoughts that allows us to do that. It allows us to think about the systems approach, systems engineering, and other related topics. In this book, we will discuss objects, namely, the aircraft and other related objects, both internal and external to the aircraft. Some of these objects are mental objects as previously discussed, for example, the specification tree.

Even at the systems thinking level, individuals should have an understanding of *holism*, the idea that everything in the entity should work together to create a whole and that emergent properties will result. According to INCOSE (2015, p. 20), emergent properties depend on the structure of the system and the relationship between the parts and with the environment. The interactions can be functional, behavioral, or performance related. The interactions can be between the parts or with the environment. Interactions can be either intentional or unintentional. Unintentional interactions can be, for example, vibrations.

The second property of systems thinking is that everything in an entity whether real or mental is connected. For real systems, the term *interact* is used. For mental systems, the term *relate* is used. So, the parts of an aircraft interact, whilst the part of a poem interrelate.

This does not mean that every part interacts with, or relates to, every other part. But it does mean that there is a chain of parts that interact or relate to each other.

The next property is synthesis, that is, the property that allows all the parts of the entity to be connected together to create a whole system. Without synthesis there would never be a whole system put together to perform the intended function. Synthesis requires the concurrent execution of both systems engineering and systems architecting, two processes that are frequently discussed separately. This book brings them together into a unified process called the *systems approach*.

Systems thinking also allows us to recognize the property of *emergence*, the property that the system has and individual parts do not have. For aircraft, the emergent property is *flight*. Flight cannot be achieved with a single engine, for example, with no wings. For a poetic system, the emergent property is the meaning of the poem itself that individual words and lines do not covey. It is logical that mental systems on the aircraft relate to each other. For example, the parts of the flight manual relate to each other and to the real parts as well. The flight manual (the mental part) will tell the pilot how to control the aircraft when a pilot tube (a physical part) has been damaged and is not functional.

Many systems contain the property of feedback loops, the property that allows the parts of the system (physical) to reinforce each other and to create balance among the parts. Aircraft control systems perform this function. This feature allows the aircraft to fly at a specific altitude and speed.

The next feature is *causality*, how things result from actions. Flight, of course, results from lift. The whole aircraft has a common cause or goal.

Finally, all of the things above result in *system mapping*. This shows an understanding of how everything in the list above relates to each other. The emergent property of flight, for example, could not happen if the aircraft were not synthesized as a whole. System mapping applies to both physical parts and mental parts and their relation to each other as illustrated above.

REFERENCES

Checkland, Peter. 1999. *Systems Thinking, Systems Practice*. New York: John Wiley & Sons.

INCOSE. 2015. *Systems Engineering Handbook*, edited by SE Handbook WorkingGroup. Seattle, CA: International Council on Systems Engineering.

4 Worldviews

Another aspect of systems thinking is the worldviews of systems. A group of INCOSE Fellows were charged with reexamining what is meant by the term system. The document (Dori et al. 2019) is a summary of that study. This team found that within the systems community the definitions of system varied widely. Figure 4.1 illustrates these views.

So the question at hand is: If the worldviews of systems specialists vary widely, how do aircraft fit into this picture? Let us first summarize these seven worldviews. Following is a summary of those views.

A worldview, from the German weltanschauung, is a mental model of reality—a comprehensive framework of ideas about the world, ourselves, and life; a system of beliefs; a system of personally customized theories about the world and how it works, with answers for a wide range of questions. In the context of systems, worldview pertains to a framework of ideas about systems. Sillitto and Dori (2017) have identified seven such principal worldviews:

Worldview 1: A formal minimalist view. This worldview consists of two or more related items, such as the hydrogen atom (two particles), the word "is," the sentence "I live," two people working together, a married couple, the earth and the moon, and a binary star. It is generally agreed that in order for a system to exist in this worldview, it must be capable of being mathematically modeled.

Worldview 2: Constructivist or a mental model. This is the worldview of Checkland (1999, p. 317), who defines a system as "a model of a whole entity." Hence, for this worldview, the earth and the moon only become a system when observed by humans. In Checkland's view, systems are intellectual constructs and not descriptions of actual real-world activity. These systems are shown in Figure 4.2.

According to Meadows (2008, p. 87), "everything we think we know about the world is a model. Our models do have a strong congruence with the real world. Our models fall short of representing the real world fully."

Worldview 3: Moderate realist. This worldview consists of both real and modeled systems. This worldview reflects systems engineering as it is often practiced in which models of real systems are developed before the real system is developed.

Worldview 4: Strong and extreme realist. This worldview maintains that only real systems are systems and they only exist in the physical world. Examples include airplanes, bridges, cars, beaver dams, and beehives as shown in Figure 4.2.

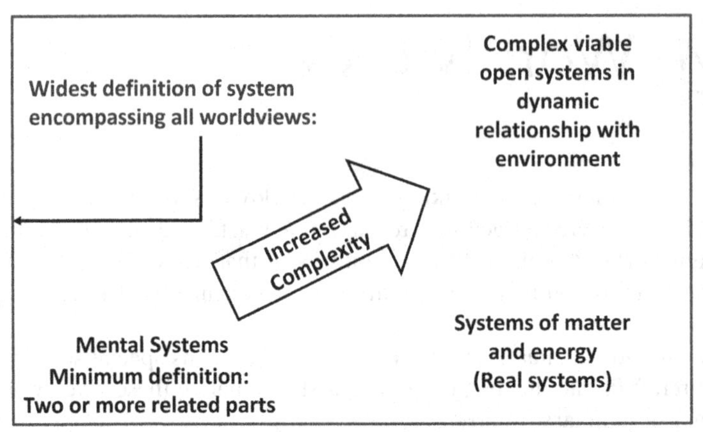

FIGURE 4.1 Worldviews.

Worldview 5: These systems consist of complex, viable, and living entities. These systems have the ability to retain system properties under stress. Among other characteristics, the following are identified: resilience, the ability to absorb and recover from major disruption; "homeostasis," the ability to maintain a condition of equilibrium within its internal environment, even when faced with external changes. When all the characteristics are viewed, the systems of this worldview are the ones that are most naturally resilient.

Worldview 6: A mode of description. This worldview states that any entity can be described as a system. In this worldview, systems still have the basic characteristics of multiple components, relationships between the components, and interactions with the outside world. However, these characteristics do not define the system; the entity is a system simply because it is

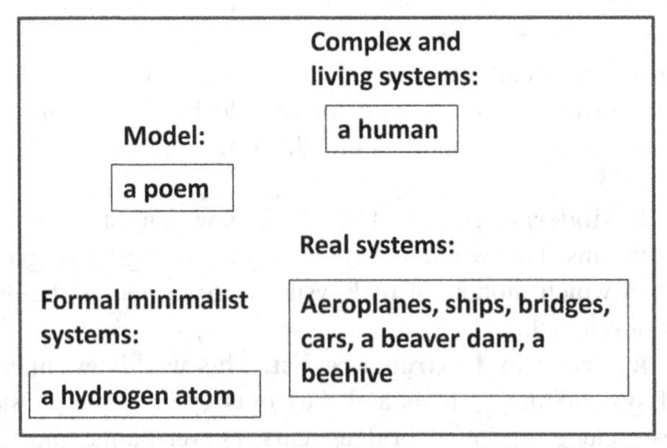

FIGURE 4.2 Examples of systems.

called a system, which is a separate ontological class. This worldview is one of the less emphasized.

Worldview 7: A process. A system in this worldview is any entity performing an activity. In this worldview a system is not defined by its structure but its ability to perform a process. Hence, the aircraft system would fall into this category not because of the number of interacting parts but rather by its ability to deliver passengers and cargo.

A VIEW OF AN AIRCRAFT ACROSS WORLDVIEWS

Now that all of the worldviews have been defined, where do aircraft fit in? Actually, they can be seen across several worldviews. They can be seen in the context of all the worldviews listed above.

Certainly, from the minimalist point of view, aircraft definitely fit. It is safe to say that all aircraft consist of two or more parts that relate to each other and exhibit emergence. Emergence in this case is powered flight.

Using the definition of a system as a "model of a whole entity" described by Checkland (1999, p. 317), an aircraft is a system. It does not matter whether the aircraft is a physical entity or not; what matters is that the anthropocentric view of the system is a model, not the physical system.

The moderate realist worldview comes into play when it is realized that an aircraft does indeed consist of physical parts, and in addition there are abstract systems within a physical system. Abstract elements consist of the myriad of "trees" that are used in the design of an aircraft. For example, there is the reliability tree, the drawing tree, and the specification tree.

The strong and extreme realist view of an aircraft is that only the physical aspects of an aircraft constitute a system. This view would, of course, eliminate the pilot.

The complex, viable, and living view of an aircraft would include the pilot. In addition many aspects of an aircraft are complex, for example, the avionics and the relation between the avionics and the pilot.

The view of an aircraft as a mode of description states that an aircraft is a system simply because humans view it as a system. It, of course, would need to satisfy some of the basic qualities of system, for example, multiple parts, interaction, and emergence.

The process view of systems fits the aircraft very easily because the aircraft has a straightforward process, namely, to carry personnel and cargo to distant destinations.

REFERENCES

Checkland, Peter. 1999. *Systems Thinking, Systems Practice*. New York: John Wiley & Sons.

Dori, Dov, Hillary G. Sillitto, Regina Griego, Dorothy McKinney, Eileen Arnold, Patrick Godfrey, James Martin, Scott Jackson, and Daniel Krob. 2019. "System Definition, System Worldviews, and Systemness Characteristics." *IEEE Systems Journal*. Preprint (accepted for publication).

Meadows, Donnella H. 2008. *Thinking in Systems*. White River Junction, VT: Chelsea Green Publishing.

Sillitto, Hillary G., and Dov Dori. 2017. "Defining 'System': A Comprehensive Approach." IS 2017, Adelaide, Australia.

5 Commercial Aircraft in the Context of Systems Theory

As Chapter 2 showed, systems engineering (SE) is a subordinate discipline of systems thinking. Furthermore, SE is the application of the systems approach to human-made systems.

The International Council on Systems Engineering (INCOSE) has updated its definition of SE. It is as follows:

> Systems engineering is a transdisciplinary approach and means, based on systems principles and concepts, to enable the realization of successful whole-system solutions. It focuses on
>
> - establishing stakeholders' purpose and success criteria, and defining actual or anticipated customer needs and required functionality early in the development cycle,
> - establishing an appropriate life cycle model and process approach considering the levels of complexity, uncertainty, and change,
> - documenting and modeling requirements for each phase of the endeavor proceeding with design synthesis and system validation, and
> - while considering the complete problem and all necessary enabling systems and services.
>
> Systems engineering provides guidance and leadership to integrate all the disciplines and specialty groups into a team effort forming an appropriately structured development process that proceeds from concept to production to operation, evolution, and eventual disposal.
>
> Systems engineering considers both the business and the technical needs of all customers with the goal of providing a quality solution that meets the needs of users and other stakeholders and is fit for the intended purpose in real-world operation.

This definition was developed by Dori et al. (2019) on behalf of INCOSE. There are several notable aspects to this definition that distinguishes it from previous definitions.

First, the term *engineering* may be interpreted in a broader sense to imply the creation of anything with the use of logic and intelligence as opposed to the traditional types of engineering, such as mechanical, electrical, or aeronautical. Jackson (2015) provides a comprehensive view of how SE is applied to commercial aircraft.

Second, the term *transdisciplinary* is introduced to apply to show how multiple disciplines can be applied together in the engineering of a system. Transdisciplinary goes beyond multidisciplinary in that it shows that cooperation among multiple disciplines is necessary to engineer a whole system. These disciplines may be more than just traditional engineering systems; they may include disciplines as diverse as psychology. For example, Billings (1997, pp. 234–246) shows how humans, for example, pilots, and automated systems should interact to control an aircraft. Billings provides a set of rules to accomplish this objective. For example, he states that "each agent [e.g., the pilot] in an intelligent human-machine system must have knowledge of the intent of the other agents [e.g., the flight control system]."

According to Meadows (2008, p. 103), "it [the system] will be sure to lead you across traditional disciplinary lines. To understand that system, you will have to be able to learn from economists and chemists and psychologists and theologians." Not to mention aerodynamicists and other engineers.

The discussion of stakeholder success criteria goes beyond immediate aircraft characteristics to discuss anticipated needs such as aircraft resilience discussed in Jackson (2015, pp. 229–243). Pariès (2011, pp. 9–27) shows how a resilient aircraft was able to recover from a bird strike. In addition, stakeholders in the commercial aircraft domain go beyond the airline customer and also include regulatory agencies, air traffic control (ATC), the flying public, and the aircraft enterprise itself.

This definition includes an understanding of complexity, which is a growing factor in modern commercial aircraft. This complexity is caused by a growing number of stochastic interfaces in subsystems, such as the avionics subsystem.

This definition emphasizes the entire life cycle of the aircraft and the ability to model the processes for each phase of the life cycle. This aspect goes beyond the conventional development phase. This definition also allows that the aircraft system will be able to *evolve* throughout the life of the system and not be constrained to a fixed set of elements.

SUMMARY OF THE EXPANDED VIEW OF SYSTEMS ENGINEERING

In conclusion, the INCOSE Fellows team had six major conclusions about the execution of SE. They are summarized here and elaborated:

First, SE should focus on information rather than processes. That is to say, system engineering has more to do with what information is passed rather than what steps are performed. The example above of sharing the intent of the various agents within a system is a good example.

SE should increase its focus on complexity. The use of model-based systems engineering (MBSE) is one approach for doing this. MBSE allows the designer to determine the stochastic interfaces within a subsystem.

SE should provide guidance and leadership to all disciplines. This effort would enhance the goal of *transdisciplinary* design. Each discipline would have to understand what its role is in relation to another discipline and how to incorporate the other discipline's parameters and goals into the whole system design.

Future SE should focus on *anticipated* stakeholder objectives and success criteria over the entire life cycle of the system.

Future SE should increase its focus on enabling systems. EIA 632 (1999) identifies the enabling system for an aircraft. They are as follows: training, maintenance, development, production, deployment, and installation; support and maintenance; and retirement or disposal. Geisert and Jackson (1998) describe the processes for the development of an aircraft maintenance system.

Systems Engineering: The Broader View

According to Sillitto et al. (2018), SE, in its modern embodiment, is a transdisciplinary and integrative approach to enable the successful realization, use, and retirement of engineered systems, using systems principles and concepts, and scientific, technological, and management methods. Jackson (2015) provides a comprehensive description of how SE is applied in a commercial aviation context.

In short, SE is the application of the systems approach to human-made systems. Commercial aircraft are an example of such a system.

Sillitto et al. (2018) have also concluded that SE is broader than traditional engineering. For example, it includes many aspects of organizational roles to accomplish its purposes. This conclusion agrees with the conclusion that engineering is "the action of working artfully to bring something about."

The current SE is implicitly a command-and-control philosophy. The new SE will employ a collaborative approach. The current SE is "ballistic" that is a trajectory set by initial conditions at the start of the life cycle. The new SE will be goal oriented, that is, it will be monitored throughout the life cycle to achieve and maintain fitness for purpose.

The current SE is defined rather in a vacuum. It is vague about the context in which it operates. The future SE is defined as a "human activity system" operating within the context in which it operates through the whole system life cycle. The future SE will be in a state of continual evolution.

The current SE is defined as a set of technical and management processes for mainly technological system development and whole life cycle support to operations. The future SE is defined as a collaboration between people and with the varied competencies needed for whole system whole life cycle success.

THE SYSTEMS ENGINEERING CONTEXT

SE can best be understood in the context in which it exists and performs. Figure 5.1 illustrates the concept of SE in context. It will be noted that Figure 5.1 does not mention systems architecting. Hence, this chart can be interpreted to apply to both SE and systems architecting, which for the purpose of system synthesis are performed together in the process we now call the systems approach.

First of all SE exists in the context of a sponsor or a problem owner. The sponsor may be a government or an enterprise. According to Sage and Armstrong (2000, p. 87), "a problem is the occurrence of an undesired aspect of the current situation that creates a gap between what is occurring and what we would like to have occur." For example, the problem may be that there is not a means to carry passengers over very long distances. The primary role of the sponsor is then to create the

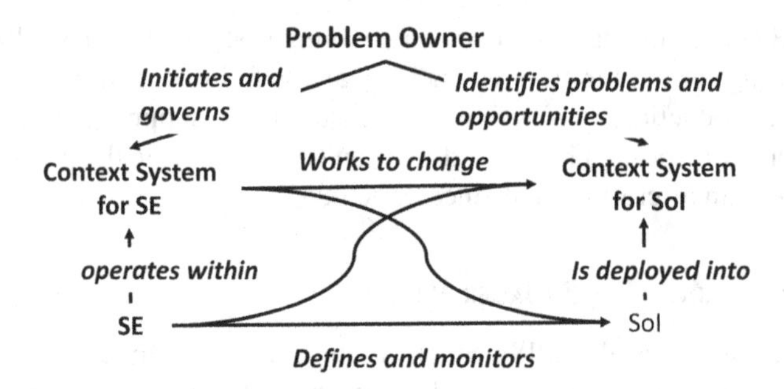

FIGURE 5.1 Systems engineering in context.

development environment in which this problem will be addressed. The sponsor also identifies the environment in which the system of interest (SOI) will operate. The SE context system is, in short, the system that conducts SE.

The SE context system creates, operates, supports, evolves, retires the SOI. According to Hitchins (2003), the context system consists of the operational environment, a threat environment, and a resource environment. The context system may also contain collaborating and competing systems. The collaborating systems would be part of a system-of-systems.

Within the context system, the system that operates within, provides information and services to changes, is changed.

The SE system architects the SOI and specifies and monitors its performance and effectiveness. The SE system observes, analyzes and understands, and monitors the system effects in the SE context. The SOI deployed, changes, is changed by the context system operational environment.

As can be seen from the above discussion, the sponsor, the SE context system, the SE system, the SOI context system, and the SOI all interact to solve the problem or opportunity identified by the sponsor.

SEVEN THOUGHT-PROVOKING IDEAS ABOUT SYSTEMS ENGINEERING

Having established that SE is a derivative discipline from systems thinking in Chapter 3, it is now necessary to relate some aspects of SE that readers may not be aware of or that may actually surprise them.

1. Idea No. 1—The first idea that may be surprising to many readers is that SE is not a branch of traditional engineering that includes, for example, mechanical engineering and electrical engineering. According to Sillitto et al. (2018), SE includes traditional engineering but also includes many aspects beyond traditional engineering. Most of these aspects are

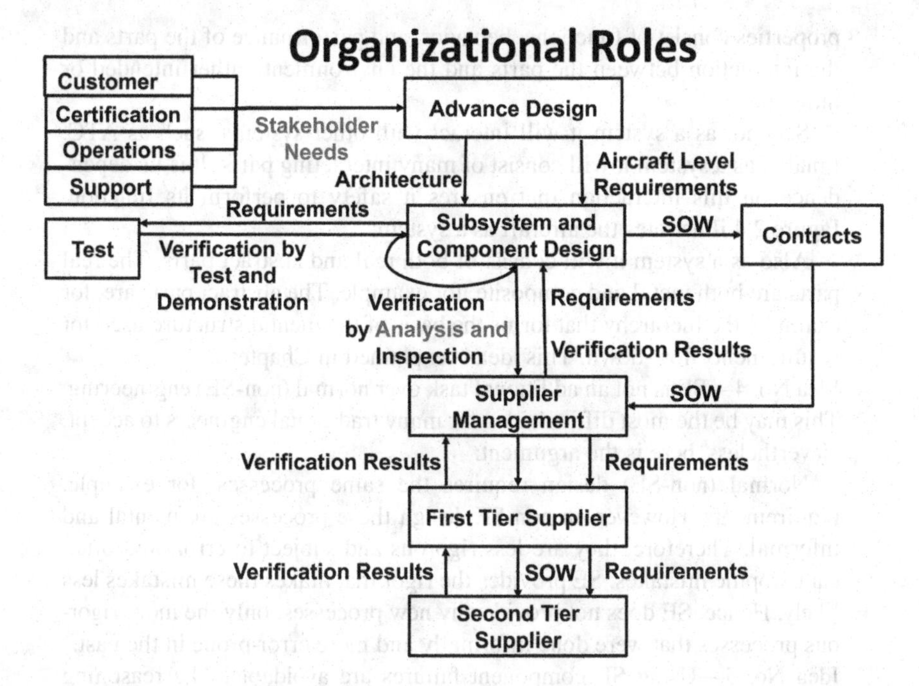

FIGURE 5.2 Organizational roles.

managerial in nature. SE would not be possible without these managerial roles. Sometimes these aspects of SE are simply called SE Management. However, most authorities include them in SE. For example, Kossiakoff and Sweet (2003) consider SE to be, for the most part, managerial.

2. Idea No. 2—The second idea is that SE does not necessarily change the organizational structure of your enterprise. Some traditional texts, for example Kossiakoff and Sweet (2003, pp. 106–108), suggest that a company's organization be structured according to the product, for example, an aircraft. However, for an existing company, this may be impractical. The important thing is that the function of each organization should be known and especially how the organizations interact with each other. Figure 5.2 illustrates how these organizations interact with respect to requirements of management and flow-down.
3. Idea No. 3—An aircraft is a system that is part of a larger system. We have explored this idea before in Chapter 2. However, it is worth emphasizing this fact for three reasons: First, as a system it will exhibit the *emergent* property of flight. According to INCOSE (2015, p. 29), the parts that interact with each other to cause emergent properties may be, for example, hardware, software, information, services, humans, organizations, and processes. Interactions may include any of the following: information, energy, and resources.

   According to INCOSE (2015, p. 29), the emergent properties depend on the parts and the relation between them and of the relationship between the whole system and the environment. The interactions that result in emergent properties consist of functions, behavior, and performance of the parts and the interaction between the parts and the environment, either intended or unintended.

Second, as a system it will interact with other systems, such as ATC. Finally, as a system it will consist of many interacting parts. It is its dependence on this interaction that ensures it safety to perform its function. Figure 2.4 illustrates the aircraft as a system.

Also as a system it will consist of both real and abstract parts. The real parts are both metal and composite, for example. The abstract parts are, for example, the hierarchy that forms the basis of the mental structure used for requirements flow-down. This idea is explained in Chapter 2.

4. Idea No. 4—SE is not an additional task over normal (non-SE) engineering. This may be the most difficult idea for many traditional engineers to accept. Nevertheless, here is the argument:

   Normal (non-SE) design requires the same processes, for example, requirements. However, in non-SE design these processes are mental and informal. Therefore, they are less rigorous and subject to error, and often catastrophic mistakes. SE provides the rigor that makes these mistakes less likely. Hence, SE does not provide any new processes, only the more rigorous processes that were done informally and more error-prone in the past.
5. Idea No. 5—Using SE component failures are avoidable. The reasoning behind this statement is that if all components are tested (or otherwise verified) to the worst-case conditions, they will never fail if conditions beyond those worst cases never occur in operation. This is one of the strongest arguments for SE.

   So when do components fail? First, they fail if the component is not tested at all. This can happen if the OEM decides, passively or actively, that testing is not required. It may be assumed tacitly that certain components will not need to be tested. If components fail this way, this may be an indication of complacency on the part of the decision makers. Or it may be a lack of knowledge on the part of the decision maker with respect to the envelope of conditions. This may have been the case with the failure of the pilot tube on Air France 447, which resulted in the loss of the aircraft and its occupants.

   The second reason for component failure is when the stress on a component exceeds the design level. This can happen knowingly as is the case for lightning strikes. The incident strike level is set very high, for example, around 200,000 amperes. It is known that this level can be exceeded, but the probability is so low that failures will be rare.

   In other cases, failures can be expected, but built-in resilience can avoid catastrophic results. Bird strike is an example. All aircraft engines are designed to withstand the impact of a bird strike of about 4 pounds. If this level is exceeded there are internal sources of power to maintain the controls of the aircraft.

   In summary, catastrophic events are very rare. This can be attributed, in part, to rigorous testing of the components and to rigorous creation of procurement specifications for those components.

6. Idea No. 6—The OEM has responsibility for product quality at all levels (the system view). This is more of a management principle than a technical principle. This goes back to the idea that an aircraft is a system. Hence, all its subsystems and components are part of that system. The system owner is the OEM. The OEM owns the whole system, the aircraft, and also the subsystems and the components.

   This principle has its greatest impact in the study of failures, not whom to blame for a failure, but rather who has the responsibility for assuring that the whole system, including subsystems and components, perform correctly under all anticipated conditions. This principle is reflected in the concept of stakeholder needs and how these needs are flowed down to all levels of the system hierarchy.

   This principle also affirms the importance of the OEM's responsibility validating the product requirements at all levels of the systems hierarchy and to validate that they have been verified in the subsystem and component designs.

7. Idea No. 7—The weakest component in the system is the human. The human is also a valuable component. We have already shown in Chapter 2 that the human is indeed a component in the SOI, the aircraft. We also showed in Idea No. 5 above that component failure is avoidable. This is much more feasible for physical components, such as hardware and software. It is far more difficult for human components.

   The Commercial Aviation Safety Team (CAST 2011) is devoted to finding solutions to the reduction of human errors in aviation accidents. CAST focuses on human error in the operational phase of an aircraft's life cycle.

   However, human error can occur at any phase of the aircraft's life cycle: conceptual design, production, operation, maintenance, and so forth. Other experts have shown that a cause of human error in some cases is *cognitive bias*. This is an error in decision making influenced by emotion and prior beliefs. This phenomenon is explored further in Chapter 7. Cognitive bias has been blamed on many major accidents.

   The seven ideas in this chapter have been presented because they may not be fully appreciated in the engineering and management community. These ideas are elaborated elsewhere in this book. More than just being surprising, it is our contention that full appreciation of these ideas will lead to higher quality aircraft.

REFERENCES

Billings, Charles. 1997. *Aviation Automation: The Search for Human-Centered Approach.* Mahwah, NJ: Lawrence Erlbaum Associates.

CAST. 2011. "The Commercial Aviation Safety Team." Last Modified 2011, accessed 16 February. http://www.cast-safety.org/about_vmg.cfm.

Dori, Dov, Hillary G. Sillitto, Regina Griego, Dorothy McKinney, Eileen Arnold, Patrick Godfrey, James Martin, Scott Jackson, and Daniel Krob. 2019. "System Definition, System Worldviews, and Systemness Characteristics." *IEEE Systems Journal* Preprint (accepted for publication).

EIA 632. 1999. *Processes for the Engineering of a System.* Arlington, VA: ANSI.

Geisert, Madrona, and Scott Jackson. 1998. *The Application of Systems Engineering to the Synthesis of Enabling Products: An Aircraft Support System.* Minneapolis, MN: INCOSE.

Hitchins, Derek. 2003. *Advanced Systems Thinking, Engineering, and Management.* Norwood, MA: Archtech House.

INCOSE. 2015. *Systems Engineering Handbook*, edited by David D. Walden, Garry J. Roedler, Kevin J. Forsberg, R. Douglas Hamelin and Thomas M Shortell. Fourth ed. San Diego, CA: INCOSE.

Jackson, Scott. 2015. *Systems Engineering for Commercial Aircraft: A Domain Specific Adaptation*, edited by Guy Loft. Second ed. Aldershot, UK: Ashgate Publishing Limited (in English and Chinese). Textbook.

Kossiakoff, Alexander, and William N Sweet. 2003. "Systems Engineering: Principles and Practice." In *Wiley Series in Systems Engineering and Management*, edited by Andrew Sage. Hoboken, NJ: John Wiley & Sons.

Meadows, Donnella H. 2008. *Thinking in Systems.* White River Junction, VT: Chelsea Green Publishing.

Pariès, Jean. 2011. "Lessons from the Hudson." In *Resilience Engineering in Practice: A Guidebook*, edited by Erik Hollnagel, Jean Pariès, David D. Woods and John Wreathhall, 9–27. Farnham, Surrey, UK: Ashgate Publishing Limited.

Sage, Andrew, and James E. Armstrong Jr. 2000. "Introduction to Systems Engineering," In *Wiley Series in Systems Engineering*, edited by Andrew P. Sage. Hoboken, NJ: John Wiley & Sons.

Sillitto, Hillary G., James Martin, Regina Griego, Dorothy McKinney, Dov Dori, Scott Jackson, Eileen Arnold, Patrick Godfrey, and Daniel Krob. 2018. "A Fresh Look at Systems Engineering – What is it, How Should it Work?" *INCOSE International Symposium* 28(1):955–970.

6 The Engineering of Systems in a Systems Theory Context

The conceptualization of an aircraft or any other large system, including the context in which it operates, must include the concept of a hierarchy as described by Bertalanffy (1968, p. 75). The hierarchy is a fundamental concept within systems theory. In addition, within aircraft development various hierarchies have been used for generations as tools for viewing different aspects of the aircraft. These hierarchies are often called "trees." There is the reliability tree, the specification tree, the drawing tree, and so forth.

These hierarchies are not, in general, physical hierarchies. Many of them are functional hierarchies as suggested by Bertalanffy. That is to say, each item in a functional hierarchy is subordinate to another item at a higher level. The aircraft itself is the highest-level item.

THE VEE MODEL

As shown in Figure 6.1, the Vee model is more than an academic artifact. On the one hand, it embodies the principles of systems thinking including hierarchy and holism. At the systems engineering level, it reflects requirements traceability, stakeholder needs, verification, and validation. This model is described in INCOSE (2015, pp. 32–36). At the same time it is a tool for guiding the development of an aircraft or any other system.

First of all, the Vee model reflects the *holism* principle of systems thinking. It shows how all the parts of a system can interact with each other to accomplish a unified purpose, which is the *emergent* property of the system. The emergent property of an aircraft is *flight*, or better put *powered flight*.

The left-hand side of the Vee is sometimes called the requirements side, but includes much more. It begins with the stakeholder needs at the top. These needs are converted into product requirements and then flowed down layer by layer through the hierarchy until the requirements for all parts have been specified. The fact that all the requirements have been specified at all levels and all requirements are derived from the higher-level part is called *traceability*.

Before the requirements can be flowed down, the *architecture* of the aircraft must be established: wings, engines, empennage, etc. This process is discussed in Chapter 8 and is therefore known as *architecting*. The stakeholder needs also drive the architecture. The hierarchy of the aircraft is part of the architecture.

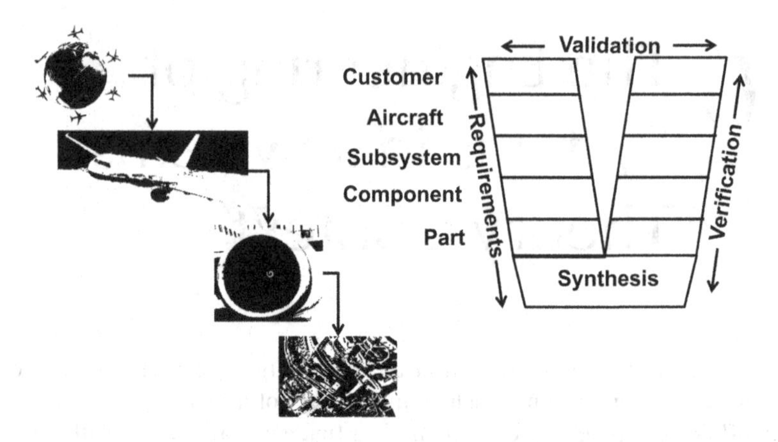

FIGURE 6.1 Vee model.

At some intermediate levels major parts will be identified, for example, the wing. All requirements for lower level parts, for example an aileron, will trace to the major part. The objective is to establish traceability from all parts to the stakeholder needs. When there are intermediate parts, the traceability will be indirect.

Hence, using the Vee model as a guide during the actual requirements development process will aid in establishing an aircraft with a holistic set of requirements.

At the bottom of the Vee, requirements will have been established for the lowest level part or subsystem. The requirements for the lowest level part or subsystem will be the result of all the higher-level requirements starting with the stakeholder needs.

However, holism is not achieved until the right side of the Vee is addressed. This side verifies the requirements for all parts and subsystems identified on the left side. At some levels, verification is achieved for entire subsystems, such as the avionics or the electrical power subsystem.

When verification is complete for all levels of the hierarchy, the final task is to determine whether the entire aircraft meets the stakeholder needs. This can be done by simulation or actual flight test. This task is called *system validation.*

THE PROCUREMENT SPECIFICATION

In modern commercial aircraft, most parts and subsystems are procured from outside companies, called suppliers. To procure these items, the aircraft company called an OEM (original equipment manufacturer) and prepared a specification to which the supplier designs and builds the part or subsystem. So, why is this specification more important than all the other specifications that might be prepared? The answer is found in the discussion of the Vee model above. It is because that specification is the embodiment of all the higher-level designs and requirements. It can be argued that if the procurement specification is correct, all the higher-level designs and requirements will be right also. And hence the whole aircraft will be right.

Most systems practitioners are familiar with these kinds of specifications and know that the primary information consists of performance requirements (of the item), constraints, and the environment in which the item is expected to operate. However, in order to meet the total goal of traceability, the specification should accomplish the following:

- All requirements in the specification should be traceable to the stakeholder needs. Stakeholders include both airline customer and regulatory agencies, operations requirements, support requirements, and the flying public.
- All requirements in the specification should be traceable to the next higher-level requirement and to all other intermediate requirements.
- Environmental requirements in the specification should include the worst-case environments, the item is expected to encounter over its lifetime, including temperature, pressure, and vibration. These environments are not only nominal in-flight environments but also abnormal conditions. These environments can be established by flight simulation.
- Requirements must include safety and cybersecurity requirements.
- The specification should reflect the full life cycle of the aircraft including retirement.
- The specification should include all constraint requirements, including EMI (electromagnetic interference) and durability. Durability is particularly important on regional aircraft, where the number of landing cycles is greater than for mainstream aircraft. Figure 6.2 shows typical durability requirements for regional jets in terms of landing cycles.
- It should contain all interface requirements, both internal and external to the aircraft.

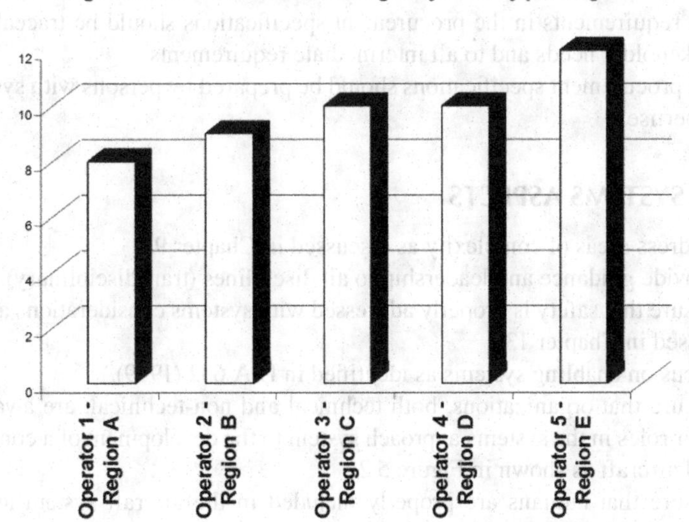

FIGURE 6.2 Durability requirements for regional jets.

- The specification should be free of task statements that belong to the supplier contract.
- The requirements in this specification should not appear in the supplier contract.
- The specification should be free of solutions. It should have requirements only. The supplier will provide the solution.
- It will contain a verification matrix to explain how each requirement is verified.
- The requirements in the specification must be achievable.
- The requirements in the specification must not be in conflict.

In addition to the procurement specification, the contract to the supplier should meet the following requirements:

- It should contain only design tasks, not solutions.
- It should not contain any requirements. They will appear in the procurement specification.
- It will acknowledge that the OEM will validate the results of the verification tasks.
- The contract will reference the procurement specification.

It is essential that this specification be prepared by a person with an in-depth knowledge of concepts like traceability and derivation.

RECOMMENDATIONS

From the above list, the two most salient recommendations are as follows:

- All requirements in the procurement specifications should be traceable to stakeholder needs and to all intermediate requirements.
- All procurement specifications should be prepared by persons with systems expertise.

OTHER SYSTEMS ASPECTS

- Address areas of complexity as discussed in Chapter 9.
- Provide guidance and leadership to all disciplines (transdisciplinary).
- Assure that safety is properly addressed with systems considerations as discussed in Chapter 13.
- Focus on enabling systems as identified in EIA 632 (1999).
- Assure that organizations, both technical and non-technical, are aware of their roles in the systems approach system to the development of a commercial aircraft as shown in Figure 5.2.
- Assure that humans are properly included in the aircraft system as discussed in Chapter 10.

- Assure that supply chain risks are properly handled as discussed in Chapter 14.
- Assure that a risk management process is in place as discussed in Chapter 11.
- Assure that methods of mitigating irrational decisions are addressed as discussed in Chapter 13.

REFERENCES

Bertalanffy, Luwig von. 1968. *General Systems Theory: Foundation, Development, Applications*. Revised ed. New York: George Baziller.

EIA 632. 1999. *Processes for the Engineering of a System*. Arlington, VA: ANSI.

INCOSE. 2015. *Systems Engineering Handbook*, edited by SE Handbook Working Group. Seattle, CA: International Council on Systems Engineering.

7 The Systems Approach

The next aspect of systems theory is the systems approach, which defines a common language and foundation for systems engineering and its other main component systems architecting. The systems approach is also described by Jackson, Hitchins, and Eisner (2010). According to Checkland (1999, p. 5), an approach "is a way of going about tackling a problem." Now that we know the basic features of systems thinking, above, we can now think about tackling problems for which systems are the solution. We don't know yet exactly what the problem is or what kind of solution is required; that is the goal of systems engineering to be discussed below.

According to Maier and Rechtin (2009, p. 8), "a systems approach is one that focuses on the system as a whole, specifically linking value judgments (what is desired) and design decisions (what is feasible). A true systems approach means that the design process includes the 'problem' as well as the solution."

According to BKCASE Editorial Board (2016) (Applying the Systems Approach), the systems approach consists of applying many processes, not the least of which are systems engineering and systems architecting. Other processes include identifying and understanding problems and solutions and synthesizing possible solutions. However, the core processes are systems engineering and systems architecting because they are the processes that lead to the actual design.

Identify Stakeholders and Formulate Agreement. In whatever domain may be of interest, every system has one or more stakeholders. A stakeholder is anyone who has an interest in the success of the system and who is affected by the system. In the aviation domain there may be many stakeholders. Agreements can be in the form of contracts or simply environmental regulations.

The primary stakeholder is, of course, the airline customer in the aviation domain. Other stakeholders include the regulatory agencies who are interested in the safety of the aircraft and its passengers and in compliance with environmental regulations, such as noise.

Other stakeholders include the aircraft crew and maintenance personnel.

Understand the Problem and Opportunities. In the aviation domain, the problem may be far greater than how just to transport passengers and cargo at a given distance. Many other problems will emerge, such as how to operate in environments of extreme heat and cold. When all of these problems are laid on top of the problem of how to perform the mission and make a profit, it is seen that the problem space is multi-dimensional.

Synthesize Possible Solutions. At this point possible solutions are notional. Is a jumbo jet required or a regional jet? All this depends on the market.

How many passengers should it carry? How many engines should it have and so forth? The developer will consult the systems architecting strategies described in the next chapter.

To define each solution, the BKCASE Editorial Board (2016) (Applying the Systems Approach) explains how systems engineering and systems architecting come together to define a common solution. For systems engineering, BKCASE identifies the three essential elements of systems engineering, namely, requirements, verification, and validation. These elements are described more fully in Chapter 6 and texts such as Sage and Armstrong (2000). For systems architecting BKCASE states that the architecture must be defined. This aspect is discussed more fully in Chapter 8 and in texts such as Rechtin (1991) and Maier and Rechtin (2009). Hence, both systems engineering and systems architecting are required to execute the systems approach.

Analyze and Select Between Alternative Solutions. At this point the considerations will become more quantitative. It should be known at this point what the range of the aircraft is, how many passengers will it carry, and how many engines will it have? Cost considerations will become dominant.

Implement and Prove a Solution. To do this, the developer will finalize a system architecture as described below and systems engineer a final solution. This solution will meet the needs of all stakeholders described above.

Deploy, Use, and Sustain a System to Solve the Identified Problem or Problems. To accomplish this goal all the systems engineering techniques and principles need to be put into full use and exercised. This step will address the design, production, operation, and maintenance throughout its entire life cycle.

RELATED DISCIPLINES

There are several other disciplines that have evolved from the Systems Approach. These will be briefly summarized here. As stated above some of these disciplines are not always practiced in the commercial aviation domain

Cybernetics. This is the study of the flow of information through a system, in our case the aircraft. In the era of fly-by-wire control, this area is extremely pertinent to commercial aviation. Cybernetics addresses how information is used to control the system through a feedback mechanism.

Operations Research. This is the methodology used almost exclusively, but not completely, in the military domain. This methodology uses mathematical models to optimize decisions. In commercial aircraft, this methodology would be most useful in the analysis of flight operations.

Systems Analysis. Once again, this methodology is widely used in the military domain. It combines *operations research, cybernetics,* and cost analysis to assist in making decisions. In the commercial aviation domain it would be most useful in the analysis of the mission of the aircraft. This methodology was developed by the Rand Corporation.

Organizational Cybernetics. This field focuses on the viability of organizations. It considers mechanisms for communication and control between the organizations. In the commercial aviation domain there are many organizations that may apply, for example, the aircraft crew, the maintenance organization, and the air traffic control organization.

Hard and Soft Systems Thinking. This is the field pioneered by Checkland (1999). It focuses on ill-defined problems in the social and political arenas. Hard systems thinking focuses on purpose, missions, and traditional engineering aspects. This is the area of systems engineering to be discussed below. Soft systems thinking focuses on complex, problematic situations.

Checkland (1999, pp. A31–A41) shows how the soft systems model (SSM) can be used to analyze the entire National Health Service (NHS) in the UK. Thus, it would be logical to conclude that it could also be used to analyze the entire national aviation system.

Note: The authors have observed that the meaning of the term *systems approach* varies among sources. It is sometimes considered equivalent to *systems engineering*. The description in this chapter is typical.

REFERENCES

BKCASE Editorial Board. 2016. "Systems Engineering Body of Knowledge (SEBoK)." Accessed 14 April. http://sebokwiki.org/wiki/Guide_to_the_Systems_Engineering_Body_of_Knowledge_(SEBoK).

Checkland, Peter. 1999. *Systems Thinking, Systems Practice*. New York: John Wiley & Sons.

Jackson, Scott, Derek Hitchins, and Howard Eisner. 2010. "What is the Systems Approach?" *INCOSE Insight*, April, 41–43.

Maier, Mark W., and Eberhardt Rechtin. 2009. *The Art of Systems Architecting*. Third ed. Boca Raton, FL: CRC Press. Original edition, 1991.

Rechtin, Eberhardt. 1991. *Systems Architecting: Creating and Building Complex Systems*. Englewood Cliffs, NJ: CRC Press.

Sage, Andrew, and James E. Armstrong Jr. 2000. "Introduction to Systems Engineering." In *Wiley Series in Systems Engineering*, edited by Andrew P. Sage. Hoboken, NJ: John Wiley & Sons.

8 Systems Architecting

Systems architecting creates the opportunity for capturing, analyzing, and understanding the stakeholder needs and understanding the problem domain, the most important point of which is to capture the underlying needs. These needs are very important because stakeholders cannot express most needs, with text or speak about the problem in a clear way, proposing solutions (tradeoffs), deriving requirements to be addressed by the system architecture and to meet the purpose or stakeholders' purposes of the system. Figure 8.1 illustrates the architecting process from the initial idea to the final realization. The primary product of the architecting process is the system architecture, both physical and functional and the relationships among the architectural components. Architecting applies at both the aircraft level and at the subsystem level.

Abstract principles form the basis for action, such as design, in many fields, such as architecture. According to Clinical Architecture (2019), the Roman architect Vitruvius articulated the following three principles of architecture:

- **Firmatis (Durability).** It should stand up robustly and remain in good condition.
- **Utilitas (Utility).** It should be useful and function well for the people using it.
- **Venustatis (Beauty).** It should delight people and raise their spirits.

None of these principles describes the exact design of any product such as an aircraft. That lack of detail is what makes them abstract. These three principles are aspects of the whole but are not parts of the whole.

The basic definition of architecting is the arrangement of the parts of a system. This arrangement can be either physical or functional. The physical arrangement pertains to, for example, the configuration of the wings and empennage. Functional arrangement pertains to, for example, how auxiliary power is provided. When these architectural decisions are made, then the systems engineering processes are used to determine such quantities as power, lift, and control. Systems engineering is dependent on the architectural arrangement. Systems engineering and systems architecting are inseparable. One cannot exist without the other.

As illustrated above, one of the methods employed by architecting is abstraction. According to Locke (1999), an abstraction is a description of the "essence" of an entity. For example, physical redundancy is the abstract description of any entity consisting of two identical branches. If an aircraft contains two identical control systems, then the abstract principle of physical redundancy will apply.

The abstract view is very important and helps to simplify the architecture by considering the essential aspects in the right moment of the process and architecting activities. Another method is the use of heuristics; it is an intuitive aspect based on

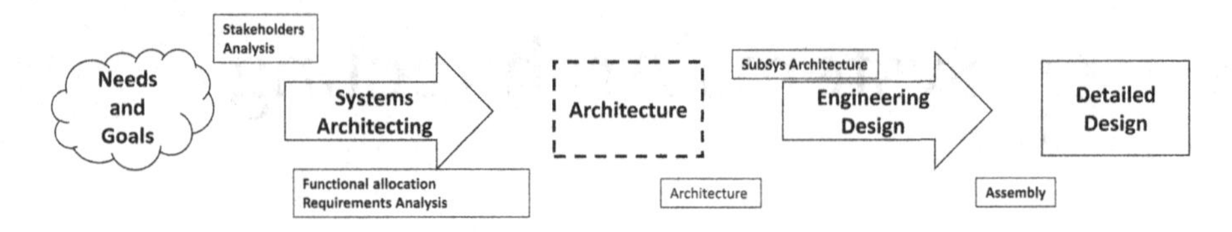

FIGURE 8.1 Architecting.

lessons learned from experience (e.g., on the job training, other development experiences and issues, etc.). The use of physical redundancy can also be considered a heuristic. Physical redundancy is but one of a set of 14 heuristics identified by Jackson and Ferris (2013).

There are different methods to use for different phases of the development process in the early phases in which the structure of information and the needs or wishes of the stakeholders are poor. At this point in the development process, the time is optimum for the architecting perspective. Following this phase, it is time to integrate the detailed information using rational and normative methods described by Maier and Rechtin (2009, p. 169). The use of abstractions and heuristics are appropriate for the early phases of concept development.

- Art: Participate (Stakeholder Based) and Heuristic (Lessons Learned).
- Normative Method: The normative method is solution based; it is more science and analytical viewpoint about the strategy solution, examples build codes (Mark Maier and Eberhardt Rechtin).
- Rational Method: The rational method is a science and analytical viewpoint about the strategy solution, example systems analysis (Mark Maier and Eberhardt Rechtin).

Art, normative, and rational methods are essential to the classic architecting approach and very useful to the original equipment manufacturer (OEM). Art is the most appropriate for systems architecting. Because of the quantitative nature, the normative and rational methods serve to introduce the systems engineering phase of development.

So, it is fair to ask if both the rational method of architecting and systems engineering are based on the principles of mathematics and science (see rational method in Maier and Rechtin (2009, p. 425)), where does architecting ends and systems engineering begins? Maier and Rechtin (2009, p. 426) answer this question by pointing out that in systems engineering decisions are based on the system as a whole. So it can be concluded that architecting is used for determining the features of a system, while systems engineering is used to specify the performance of the entire system.

Finally, systems architecting is a process driven by a client's purpose or purposes. Clearly, if a system is to succeed, it must satisfy a useful purpose at an affordable cost for an acceptable period. According to Maier and Rechtin (2009, p. 260), "success is defined by the beholder, not the architect."

In the INCOSE references, ISO/IEC/IEEE 42010 Systems and Software Engineering, Architecture description (ISO 2011) provides a useful description of the architecture considering the stakeholder concerns, architecture viewpoints, architecture views, architecture models, architecture descriptions, and architecting throughout the life cycle (BKCASE Editorial Board 2016).

The life cycle process is very useful in the commercial aircraft domain in which the aircraft is considered as a system. how this objective and the title of this book, a lot of consideration of the architecting process and activities need to be addressed to help understanding the cost and the correct context and environment of the whole solution are.

The architecting process applies to all levels of the aircraft hierarchy as illustrated in Figure 4.1. It is therefore integral to the systems view of an aircraft. Architecting receives the most attention at the top of the aircraft hierarchy, defining the relations among the fuselage, wings, engines, and the empennage. However, aircraft contain within their boundaries many distributed subsystems. These include the control system, the hydraulic system, the environmental control system (ECS), the fuel system, and the power system. All of these subsystems are subject to the architecting process. In short, the architecting process covers the entire aircraft.

As an example of architecting at the subsystem level, take the ECS for example. This subsystem consists of many sensors scattered throughout the aircraft. The location of the individual sensors is the subject of systems architecting at the subsystem level.

SYSTEMS ARCHITECTING AND SYSTEMS ENGINEERING

Figure 2.4 shows that both systems architecting and systems engineering are subordinate disciplines to the systems approach, which is the subject of this book. So, if this true, what are the differences and how do they interact?

The general view is that systems architecting is more of an art as explained by Maier and Rechtin (2009, p. 169). On the other hand, systems engineering is regarded as more of a quantitative science. Hence, systems engineering is more of a left-brain discipline, while systems architecting is more of a right-brain discipline. For, example, systems architecting is the discipline responsible for concept development, which according to Maier and Rechtin is an artistic endeavor. On the other hand, systems engineering is responsible for the achievement of system performance, a right-brain activity.

According to Britannica (2019), the idea that people are either right-brained or left-brained is untrue and a total myth. Hence, anyone can be right-brained or left-brained and hence either a systems engineer or a systems architect, or both. People who are both analytic and artistic are sometimes called whole-brained.

In other words, the idea that systems architecting and systems engineering have to be performed by different people has no support in the literature. They can be performed by the same person, which is an advantage in itself.

Furthermore, in order to define and develop a whole system, both systems architecting and systems engineering must be employed. And if this can be done by a single person, it is much better.

According to BKCASE Editorial Board (2016) (Applying the Systems Approach), both systems engineering and systems architecting are two of the processes required to define a system through the systems approach among other processes.

SYSTEMS ARCHITECTING IN THE AIRCRAFT LIFE CYCLE

The INCOSE (2015, p. 29) describes the aircraft development phase from the point of view of the "typical high-tech commercial systems integrator," which is normally the OEM. The initial study period is divided into three phases: The first is the "user requirements definition phase," which is often called the "user needs" phase. This phase does not determine the design of the aircraft.

The second phase is the concept definition phase. This is the phase in which systems architecting is the dominant activity. Not all architecting occurs at the aircraft level. The major subsystems have their own architectures that have to be determined in the proper phase, normally following the aircraft-level architecting.

The third phase is the Aircraft-Level Requirements phase. In this phase the requirements for the specific aircraft systems are specified.

These three phases are not necessarily chronological since architecting occurs at all levels. Furthermore, these phases are intermingled with the aircraft hierarchy itself that was described in Chapter 4. This is all in agreement with the systems principle of hierarchy.

Figure 8.2 illustrates the interrelation between the systems architecting and systems engineering in the concept development phase of a program.

User Requirements Definition Phase. In the first phase the role of the architect is to reduce ambiguity and to elicit needs and try to keep the architecture as solution neutral as possible. This phase is often called the "User Needs phase." This name emphasizes the fact that the requirements identified in this phase are not design requirements prioritizing high-level requirements that result from the goals; the requirement is more related to what my system will do to meet the user need. The needs of the stakeholders and beneficiaries guide the entire process; every decision should be traced back to them. The requirements identified in this phase are tradeable throughout the development process. So what is more important to the user,

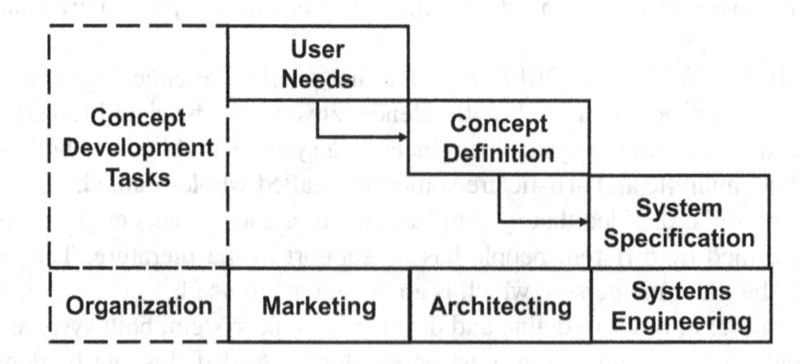

FIGURE 8.2 Concept development.

range or speed? The answer to this question, and many others, will have a strong influence on the aircraft concept.

Aircraft-Level Architecting Phase. In INCOSE (2015, p. 29), this phase is called the Concept Definition Phase. However, Aircraft-Level Architecting is a more exact description of what happens in this phase. For example, the role of the architect is to come up with a set of possible concepts and develop some of them. So, this phase includes the development of the highest level of the architecture, encompassing the system in its surrounding context within a broader system. This phase may also include a few trade-studies to assist in the definition.

Aircraft-Level Requirements Phase. INCOSE (2015, p. 29) refers to this phase as the System Definition phase. In this phase the aircraft architecture is decomposed to the aircraft elements (avionics, propulsion, and so forth) specifying them both in form and function. In terms of requirements, the functional description of the architecture will be completed when we specify the functions and allocate the requirements, proceed with a functional analysis. Also it will be remembered that the individual aircraft systems will have architectures of their own, for example, the fuel system architecture. Hence, at the end of phase 3 both the aircraft architecture and element requirements will be defined.

The purpose of the above discussion was to show how the aircraft and element architectures are developed using both systems architecting and systems engineering throughout the development phase of the aircraft. In addition, it showed how two subordinate disciplines of systems science enable this development.

AN ARCHITECTING TOOL: THE RICH PICTURE

A tool used in the architecting process is the "rich picture" as shown in Figure 8.3. This tool is a method of seeing in a very informal way the relation between the parts of the system, their function, and the options available to the architect.

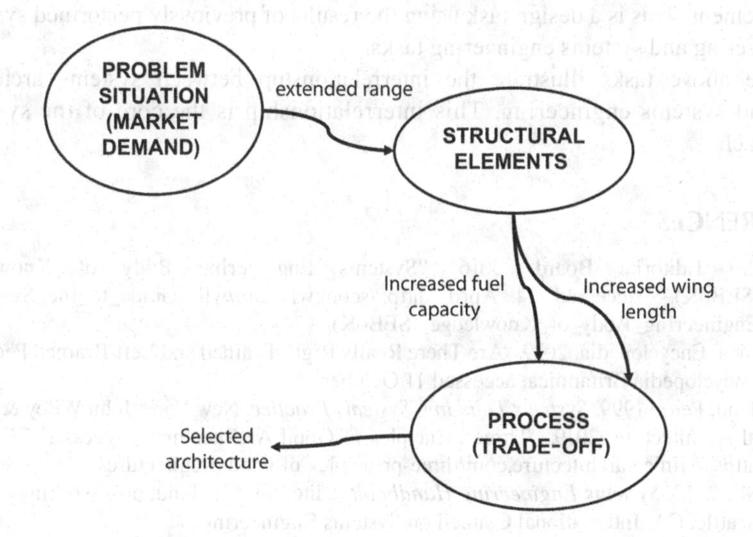

FIGURE 8.3 The rich picture.

The rich picture is suggested by Checkland (1999, p. 317) who says that it is "an expression of the *problem situation* compiled by an investigator, often by examining elements of the *structure*, elements of *process*, and situation *climate*" [italics as quoted].

ARCHITECTING: AN OVERVIEW

For a more comprehensive view of the architecting process, an overview tasks below guides the reader on the perception about the knowledge that involved on this process.

Task 1: Solution definition. This task should include an analysis of mission objectives, market objectives, and customer objectives (same place in the process, three different contexts).

Task 2: Life cycle process analysis. This task includes the identification and analysis of all system life cycle processes.

Task 3: Stakeholder analysis. This task includes the identification of stakeholders and stakeholder concerns.

Task 4: Functional analysis and architectural design. This task establishes the role of functional analysis in design.

Task 5: Functional design. This task performs functional decomposition aiming at allocable functions.

Task 6: Architectural design. This task performs functional and performance allocation. This task requires both systems engineering and systems architecting.

Task 7: Trade-off and selection. This task performs technical evaluation of alternatives. This is primarily a systems engineering task.

Task 8: Interface design. This task determines interface types. This is primarily a systems architecting task.

Task 9: Detailed design. This task performs a full characterization of each system element. This is a design task using the results of previously performed systems architecting and systems engineering tasks.

The above tasks illustrate the interrelationship between systems architecting and systems engineering. This interrelationship is the core of the systems approach.

REFERENCES

BKCASE Editorial Board. 2016. "Systems Engineering Body of Knowledge (SEBoK)." Accessed 14 April. http://sebokwiki.org/wiki/Guide_to_the_Systems_Engineering_Body_of_Knowledge_(SEBoK).

Britannica, Encyclopedia. 2019. "Are There Really Right-Brained and Left-Brained People?" Encyclopedia Britannica, accessed 11 October.

Checkland, Peter. 1999. *Systems Thinking, Systems Practice*. New York: John Wiley & Sons.

Clinical Architecture. 2019. "Three Principles of Good Architecture." Accessed 27 April. https://clinicalarchitecture.com/three-principles-of-good-architecture/

INCOSE. 2015. *Systems Engineering Handbook*, edited by SE Handbook Working Group. Seattle, CA: International Council on Systems Engineering.

Jackson, Scott, and Timothy Ferris. 2013. "Resilience Principles for Engineered Systems." *Systems Engineering* 16(2):152–164.

Locke, John. 1999. "An Essay Concerning Human Understanding." Pennsylvania State University, accessed 8 February. ftp://ftp.dca.fee.unicamp.br/pub/docs/gudwin/ia005/humanund.pdf.

Maier, Mark W., and Eberhardt Rechtin. 2009. *The Art of Systems Architecting*. Third ed. Boca Raton, FL: CRC Press. Original edition, 1991.

9 Complexity in a Systems Theory Context

In recent years commercial aircraft have become more complex meaning that there are many components that interact with each other and consist of humans. This fact leads to the conclusion that the management of complexity is a high priority in the development of commercial aircraft.

According to Hybertson and Sheard (2008, pp. 13–16), systems engineering and by extension the systems approach, is "currently experiencing a significant expansion of scope beyond its comfort zone. This expansion is predominantly in the general area of '*complex systems*' by which we mean systems that behave more like organisms and less like machines."

In addition, Warfield (2008) encourages systems engineers to move toward a systems science platform for systems engineers. Complexity is one issue for which systems science can be a benefit.

For a comprehensive look at what complexity means, Sillitto et al. (2019) provide the following definition for a complex system:

> A complex system is a system in which there are nontrivial relationships between cause and effect: each effect may be due to multiple causes; each cause may contribute to multiple effects; causes and effects may be related as feedback loops, both positive and negative; and cause-effect chains are cyclic and highly entangled rather than linear and separable

Other researchers have provided other perspectives of complexity. For example, Thurner, Hanel, and Klimek (2018, p. 7) state that complex systems can be characterized as an extension to physics in the following ways:

- Complex systems are composed of many elements, components, or particles. These elements are typically described by their state, velocity, position, age, spin, color, wealth, mass, shape, and so on. Elements may have stochastic components.
- Elements are not limited to physical forms of matter; anything that can interact and be described by states can be seen as generalized matter.
- Interactions between elements may be specific. Who interacts with whom, when, and in what form is described by interaction networks?
- Interactions are not limited to the four fundamental forces but can be of a complicated type. Generalized interactions are not limited to the exchange of gauge bosons, but can be mediated through exchange of messages, objects, gifts, information, even bullets, and so on.

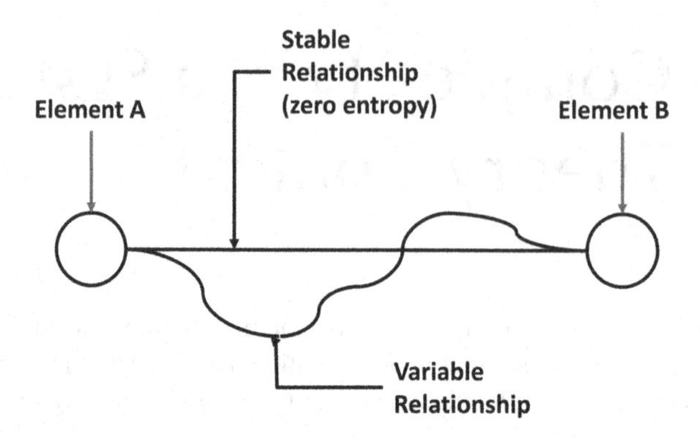

FIGURE 9.1 Entropy variation between elements.

From the above it can be concluded that time-varying and stochastic interactions are a dominant feature of complex systems. These interactions may be between humans and machines or among physical elements of the aircraft.

One of the properties of complexity according to some researchers, for example Marczyk and Deshpande (2006), is the entropy that results from the stochastic interactions between elements. Figure 9.1 shows how the entropy can vary between elements.

Complexity and Architecting

The architecting process normally faces the phenomenon of complexity. There are various techniques to manage complexity in the architecting process to provide the best and adequate solution to the stakeholders. In this process, the designer has the opportunity to define the requirements of the elements of the system architecture and the interfaces between these elements. These interfaces are a primary aspect of complexity and the architecting process is the process and the momentum that exercise to manage it with the creative's solutions and approaches.

As explained in Chapter 8, managing complexity is one of the key aspects of the art of systems architecting. Establishing clean interfaces, which minimize interaction between components, is a critical skill in the architecting process.

The definition of complexity provides a clue as to how complexity can be managed. Maier and Rechtin (2009, p. 424) state that complexity is "a measure of the numbers and types of interrelationships among system elements." So *numbers* can refer both to the number of elements and the number of interfaces.

Second, one of the heuristics quoted by Maier and Rechtin (2009, p. 397) is that "complex systems will develop and evolve within an overall architecture much more rapidly if there are stable intermediate forms than if there are not." The latter can be summarized by saying that during design interfaces should be made as stable as possible, that is to say, as deterministic as possible and not stochastic. Some authorities, for example Marczyk and Deshpande (2006), have suggested

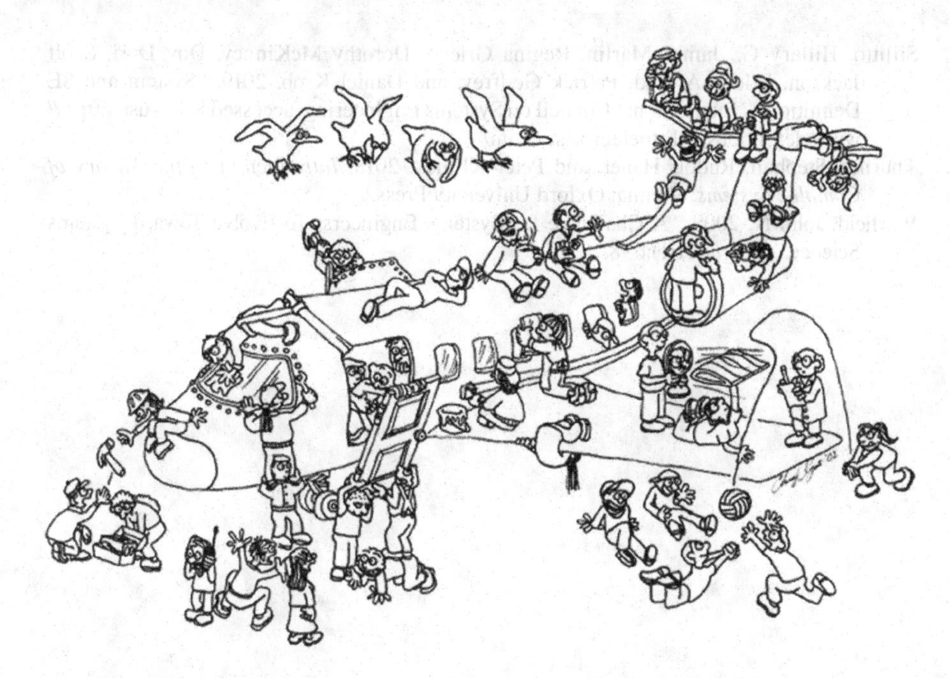

FIGURE 9.2 Complex airplane.

(Used with permission by the New England Complex Systems Institute.)

that the probabilistic nature of interfaces can be measured by Shannon entropy (Shannon 1948).

In summary, as a simple guide to the design of a system or subsystem, the following three rules should be followed:

1. The design should have as few components as possible.
2. The design should have as few interfaces as possible.
3. The interfaces between the components should be as deterministic as possible.

Figure 9.2 is a depiction of a complex airplane by artist Cherry Ogata of Japan.

REFERENCES

Hybertson, Duane, and Sarah Sheard. 2008. "Integrating Old and New Systems Engineering Elements." *Insight* 11(1):13–16.

Maier, Mark W., and Eberhardt Rechtin. 2009. *The Art of Systems Architecting.* Third ed. Boca Raton, FL: CRC Press. Original edition, 1991.

Marczyk, Jacek, and B. R. Deshpande. 2006. "Measuring and Tracking Complexity." Conference Paper Presented at International Conference of Complex Systems, Boston MA. June 2006.

Shannon, C. E. 1948. "A Mathematical Theory of Communication." *The Bell System Technical Journal* 27(3):379–423.

Sillitto, Hillary G., James Martin, Regina Griego, Dorothy McKinney, Dov Dori, Scott Jackson, Eileen Arnold, Patrick Godfrey, and Daniel Krob. 2019. "System and SE Definitions." International Council on Systems Engineering, accessed 8 August. https://www.definitions.sillittoenterprises.com/

Thurner, Stephan, Rudolf Hanel, and Peter Klimek. 2018. *Introduction to the Theory of Complex Systems*. Vienna: Oxford University Press.

Warfield, John N. 2008. "A Challenge for Systems Engineers: To Evolve Toward Systems Science." *Insight* 11(1):6–8.

10 Humans in the System

Humans in aircraft systems are a two-edged sword. On the positive side, they provide the capability of recognizing adversities and responding to them in an intelligent way. On the negative side, they provide the source of human error generally known as cognitive bias, which is a mistake caused by previous beliefs, emotion, context, or other stress factors.

Chapter 5 showed that there are seven worldviews of systems, two of which are relevant to the topic of humans and their relations to systems. One worldview is the extreme realist view in which real systems consist of matter and energy. In this worldview, the human is not part of the system but rather interfaces with the system, as a pilot does.

The other worldview of interest with respect to humans is the complex and viable systems worldview. Although this worldview may include other system types, humans would be a logical type within this worldview. There is general agreement that humans are indeed complex systems themselves and can make both wise decisions and unwise errors.

On the complex side, humans perform a vital role in the resilience of systems, for example, aircraft systems. According to Jackson and Ferris (2013), one of the most important resilience principles is the *human in the loop* principle that states there need to be humans in the system when there is a need for human cognition. Apollo 11 is the best example of the *human in the loop* principle successfully applied.

Another aspect of humans in the aircraft system is the interaction between the human and the aircraft flight control systems. Billings (1997) lays out the requirements for this interaction in a direct and logical way.

Humans are the central source of errors caused by cognitive bias. Kahneman (2011) and Thaler and Sunstein (2008) have many of these biases that have been shown to result in decision errors, many of them catastrophic. They have also pointed the way to techniques that can be used to influence the decisions for more favorable choices.

HUMANS IN AN AIRCRAFT RESILIENCE CONTEXT

According to BKCASE Editorial Board (2016), resilience is "the ability to maintain capability in the face of adversity." In general, this definition is often interpreted to mean that a system should be able to recover from an adversity when it is disrupted in any way. However, recovery does not mean that the system should be able to restore its original performance before the adversity was encountered. It only means that it should be able to maintain a level of capability that is expected by the system owner.

Resilience is broader than safety. Safety focuses on the protection from loss of life or property. Resilience focuses on maintaining capability. Resilience is not a specific property built into modern aircraft although most aircraft have many resilience capabilities. One of these will be discussed below.

Jackson and Ferris (2013) have identified 14 top-level principles that have been shown to enhance the resilience of engineered systems, such as aircraft. Most of these principles have a direct relation to the physical architecture (design) of any system. For example, the *physical redundancy* principle states that a system should be designed with two or more identical branches. This is why most aircraft have multiple engines. The use of these principles is an example of system architecting discussed in Chapter 8.

A more relevant principle for this section is the *human in the loop* principle. This principle calls for human cognition to be incorporated into the system when needed. This principle was a major factor in the well-known US Airways Flight 1549 case study also known as the Miracle on the Hudson. This case described by Pariès (2011) explains how an aircraft suffered a bird strike upon take-off from an airport in New York. This bird strike caused both engines to fail thus depriving the aircraft of internal power with which to control itself. This is where the *human in the loop* principle comes in.

The *human in the loop* principle becomes part of the *functional redundancy* principle described by Jackson and Ferris (2013). This principle calls for two different ways to maintain capability. The first way was the *absorption* principle attempted by the engines, which failed. The second way consisted of both internal power and control. Internal power was provided by the ram air turbines (RATs). Control was provided by the pilot, as the *human in the loop.*

The net result was that the aircraft was able to ditch in the Hudson River saving all 155 persons on board.

HUMANS IN AVIATION AUTOMATION CONTEXT

A necessary human task is to control the aircraft, even if automated systems are involved in the control. Billings (1997) has identified a set of rules for determining how the human and the automated system should interact with each other. The purpose of these rules is to keep the human from making a mistake, which has been done. Hence, this is another example of the *human in the loop* principle.

Figure 10.1 shows an interesting question between automation and the human's level of awareness about it. The key point of the graph is the representation of an automation in operation in relation to the advancement of time and the level of automation putting the human being in system monitoring. In this way, it creates a human machine cognitive context delta of understanding the system. As we can see in the figure, this delta shows that the setting or context of machine reading and understanding of the machine is not the same as the human being in the conception of design, so when we have possible failures (point in red in the figure) in automatism there is no transition from machine to human context and its increasing in time; monitoring is an interpretation of indicators of what the machine is doing, but usually when failure of automation occurs, the human does not always understand what the problem is. This scenario can lead to many wrong decisions being made because of a misunderstanding of the current system context and state. This transition point needs to be well taken care of in design, especially in commercial aircraft due to the high degree of automatism.

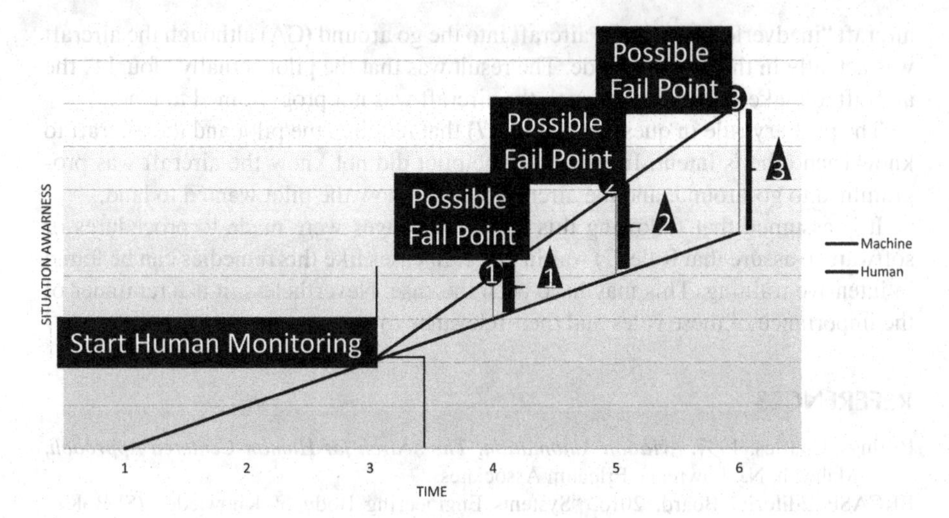

FIGURE 10.1 Humans and machine learning.

Of all these rules, the one that stands out is that "each agent [e.g., the pilot or the machine] in an intelligent human-machine system must [underlining added] have knowledge of the intent of the other agent."

The Billings Rules

One factor in human error is the relation between the human operator and the automated system. Failure to comply with these rules may be blamed for at least one major catastrophic accident. Billings (1997, pp. 232–246) calls these rules requirements, but they can also be called heuristics since they are simply common-sense rules based on the author's experience. Following are the primary rules; the reader is referred to the text to understand more details about each rule:

1. The human operator must be in command.
2. To command effectively the human operator must be involved.
3. To remain involved the human operator must be appropriately informed.
4. The human operator must be informed about automated systems behavior.
5. Automated systems must be predictable.
6. Automated systems must also monitor the human operator.
7. Each agent in an intelligent human-machine system must have knowledge of the intent of the other agents.
8. Functions should be automated only if there is good reason for doing so.
9. Automated systems should be designed to be simple to train, learn, and operate.

Most of these rules are self-evident and logical. However, there is at least one incident in which one or more of these rules were ignored. We would direct your attention to the Nagoya accident of 1994 as documented by Ladkin (1996), the pilot of this aircraft "inadvertently" put the aircraft into the go around (GA) although the aircraft was actually in the landing mode. The result was that the pilot actually "fought" the aircraft to make it land even though the aircraft was not programmed to land.

The primary rule in question is Rule (7) that requires the pilot and the aircraft to know each other's intent. In this case, the pilot did not know the aircraft was programmed to go-around, and the aircraft did not know the pilot wanted to land.

It is assumed that following this incident changes were made to procedures or software to assure that Rule (7) was in place. In cases like this remedies can be found in intensive training. This may have been the case. Nevertheless, it is a reminder of the importance of these rules and their relevance to human error.

REFERENCES

Billings, Charles. 1997. *Aviation Automation: The Search for Human-Centered Approach.* Mahwah, NJ: Lawrence Erlbaum Associates.

BKCASE Editorial Board. 2016. "Systems Engineering Body of Knowledge (SEBoK)." Accessed 15 April. http://sebokwiki.org/wiki/Guide_to_the_Systems_Engineering_Body_of_Knowledge_(SEBoK).

Jackson, Scott, and Timothy Ferris. 2013. "Resilience Principles for Engineered Systems." *Systems Engineering* 16(2):152–164.

Kahneman, Daniel. 2011. *Thinking Fast and Slow.* New York: Farrar, Straus, and Giroux.

Ladkin, Peter B. 1996. Resume of the Final Report of the Aircraft Accident Investigation Committee into the 26 April 1994 crash of a China Air A300B4-622R at Nagoya Airport, Japan. In *The Nagoya A300-600 crash.* Bielefeld UK: University of Bielefeld - Faculty of Technology.

Pariès, Jean. 2011. "Lessons from the Hudson." In *Resilience Engineering in Practice: A Guidebook*, edited by Erik Hollnagel, Jean Pariès, David D. Woods, and John Wreathhall, 9–27. Farnham, Surrey: Ashgate Publishing Limited.

Thaler, Richard H., and Cass R. Sunstein. 2008. *Nudge: Improving Decisions About Health, Wealth, and Happiness.* New York: Penguin Books.

11 Risk

BASIC RISK THEORY

The basic idea behind risk theory is that systems will incur anomalies that will have cost, technical, and schedule consequences as a result of prior actions, events, or conditions. The actions may be human stimulated, such an incorrect navigation action or they may be external such as abnormally bad weather.

Cost, technical, and schedule consequences may be interrelated. For example, a technical consequence, such an engine failure, may have both cost and schedule consequences. Likewise, a managerial consequence, such as the selection of an unqualified supplier, may have technical consequences, such as an underperforming component. Therefore, it is not valid to separate cost, technical, and schedule risks into separate and independent risks.

Risks are evaluated using the chart shown in Figure 11.1. This chart consists of two axes: the likelihood axis and the consequence axis. The likelihood is estimated on the history of similar incidents. The likelihood is based on the probability that such an event will occur *if nothing is done to prevent it from happening.* It is not valid to say that we would never let that happen. That is the sort of thing that can be said when mitigation approaches are considered.

Although the likelihood scale in this figure is qualitative, the best practice should be as quantitative as possible. For example, if a supplier has a history of delivering a product late 50% of the time, a likelihood of moderate would be appropriate.

Regarding consequence, you can create your own scale. For example, a catastrophic disaster would be considered High on the consequence scale. A degradation in performance would be at the appropriate level below that.

So what is this chart used for? Traditionally, it is used to determine the type of handling approach to apply to the risk. Conventionally, only risks in the medium and high levels are worth any handling at all.

RISK HANDLING

Traditionally there are four methods of handling risks. Conrow (2003, pp. 365–387) provides a comprehensive discussion of risk handling. Following is a summary of the principal techniques:

- **Assumption.** This method concludes that the risk at hand is worth accepting provided the consequences are not too severe. The cost of risk assumption can be handled by management reserve, if it exists.

  A common risk in commercial aircraft is the risk of overweight, especially for the first plane delivered. This risk is so common that it is expected that there will be a management reserve to cover it. This management reserve will go toward conducting a weight-reduction program to bring the weight down.
- **Control.** This method, sometimes called mitigation, consists of taking steps either to reduce the probability of occurrence or the level of consequence itself.

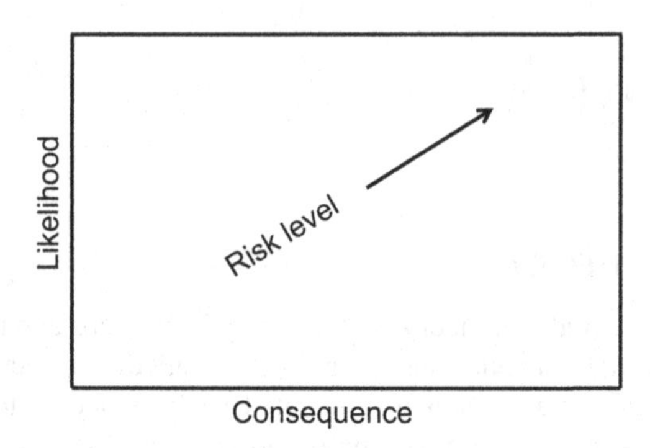

Adapted from Jackson (1997)(p. 134) and Jackson (2015)(p. 217)

FIGURE 11.1 Risk.

The obvious control techniques are schedule slip or increase in resources or personnel. These choices can be avoided if the program is well planned. Another control option is redesign; this may be an option if the component does not meet its performance requirements. Of course this will have schedule and cost consequences.

In the case of the supplier management risk discussed in the Supplier Management section, if the cost overruns are not accounted for in the budget, the OEM would normally have to employ the control method to make up for any performance, cost, or schedule shortfalls that this risk may cause.

In general this method can be avoided by good management and planning. It is expected to be necessary only in the case of surprises but possible consequences.

- **Transfer.** This method is used normally when a risk is transferred from one organization to another. However, it should only happen when both organizations are agreeable. The two organizations of importance are the OEM and the customer. In addition these risks should be accounted for in the contract with the customer. A possible performance shortfall of the aircraft would normally be accounted for this way.

  If the performance shortfall is not accounted for in the contract, then the control methods would most likely be called for.
- **Avoidance.** This is the term to represent the case in which a requirement or a design is changed before the consequence is incurred. The requirements may be changed in the procurement specification discussed in the Procurement Specification section before the supplier has actually created hardware or software.

SOME FINAL COMMENTS ON RISK

This section might be called Avoiding the Logical Pitfalls of Thinking about Risk. These comments are based on real incidents that haven been observed in industry.

The first pitfall is thinking that risk is a sign of bad management. This is not the case. Most risks are from outside sources. The comment is heard, "I don't have risk;

I am a good manager." It is true that some risks may result from management practice. Take, for example, the *optimism* bias discussed in the Humans section. This risk occurs as a result that managers, or anyone else, have an overly optimist view of the success of the system.

But for the most part, risks are externally generated, such as supplier designed components.

The next problem is that many risk analysts go to great lengths to avoid putting risks on the risk list. For example, they say that is something they deal with anyhow. The response is so what? If problems keep resulting in cost and schedule consequences in spite of all the efforts, it should go on the list.

Another excuse is that it is a customer item. If so, assign the Transfer method and discuss it with the customer.

In short, put it on the list, analyze it, and if it is an Assumption item, the job will have been done.

RECOMMENDED ACTIONS

The first recommended action is to establish a risk management process. The Systems Engineering organization is a logical home for this process. This process will contain a Risk Review Board.

The second recommended action is to take steps to assure that risk management is *independent*. There are two aspects to independence: First, the Risk Review Board must be organizationally independent of the program in question. It can either be part of another program or another organization. But it cannot be part of the program in question. Persons on the Risk Review Board cannot be part of the program in question.

Second, it must be financially independent. That is to say, the Risk Review Board cannot receive funds in any way from the program in question.

Finally, the Risk Review Board's role will be only advisory. It will not have veto power over any program actions, such as the selection of a supplier. However, its influence should be substantial.

REFERENCE

Conrow, Edmund H. 2003. *Effective Risk Management: Some Keys to Success*. Second ed. Reston, VA: American Institute of Aeronautics and Astronautics.

12 Cybersecurity

This chapter addresses the external and internal cybersecurity threats affecting commercial aircraft and the approach to dealing with those threats. This topic exists within the systems principle described in Chapter 4 that a system, the aircraft, exists within an environment. The cybersecurity environment is one that aircraft developers must deal with. The aircraft is vulnerable to threats from any system the aircraft is connected to, for example, maintenance.

Use of computer networks and loadable software enables time-efficiency and cost reduction during aircraft life cycle when compared to legacy solutions. As aircraft becomes connected to an online environment, it is susceptible to security threats impacting safety as described in RTCA (2014b) for business and operations.

These security threats due to intentional or unintentional electronic interaction can reduce safety levels and impact operational and business aspects such as passenger confidence, airline reputation, and processes (e.g., flight delays). This section presents a cybersecurity risk assessment conducted on in-service aircraft, through an application of tailored in accordance with RTCA (2014b) guideline considering aircraft embedded systems and exercising threat scenarios to identify sources of threats to safe operation, providing insights for strengthening confidentiality, integrity and availability of assets to improve cybersecurity aspects in accordance with FAA (2019b).

The aircraft's new electronic architecture provides interfaces allowing connection to airplane electronic systems and networks, and access from aircraft external sources (e.g., operator networks, wireless devices, internet connectivity, service provider satellite communications, electronic flight bags, USB drives, SD cards, etc.) to the previously isolated airplane electronic assets. This new design may result in network security vulnerabilities from intentional or unintentional corruption of data and systems required for the safety, operations, and maintenance of the airplane (EASA 2016). The architecting of the electronic system follows the principles of Chapter 8. It will be noted that the architecting of the electrical system falls under the architecting of subsystems described in Chapter 8.

The regulatory document (FAA 2019a) has assigned the Aviation Rulemaking Advisory Committee (ARAC) a new task to provide recommendations regarding Aircraft Systems Information Security/Protection (ASISP) rulemaking, policy, and guidance on best practices for airplanes and rotorcraft, including both certification and continued airworthiness. The FAA has issued policy statement (FAA 2014) that describes when the issuance or special conditions are required for certain aircraft designs.

Eurocae and WG-72 (2014) are developing airworthiness security guidelines and security objectives based on the ARAC ASISP working group report.

The FAA (2014), issued in August 06, 2014, provides resource for Airworthiness Authorities (AA) and the aviation industry for certification when the development or modification of aircraft systems and the effects of intentional unauthorized electronic interaction can affect aircraft safety. RTCA (2014a) provides a set of methods and guidelines that may be used within the airworthiness security process including cybersecurity.

AIRWORTHINESS CERTIFICATION PROCESS OVERVIEW

Airworthiness Security Process overview starts from a new development or existing product modification. Figure 12.1 presents the main activities of Airworthiness Certification Process (aircraft and/or system level). Plan for Security Aspects of Certification (PSecAC) describes the security activities to be performed, and the PSecAC Summary is released at end of the security activities to record the performed activities, any deviation, and the generated artifacts (RTCA 2014b).

AIRCRAFT EVOLUTION AND CONNECTIVITY ASPECTS

In this context, in-service aircraft designs approved by Authorities prior to 2008 (EASA 2016) had no analysis required by Authorities to be conducted in cybersecurity area.

On a more connected world, connectivity has become mandatory to reduce operational costs and increase passengers' comfort. These characteristics aim to support airliners to download and distribute data between aircraft and back office operations faster than traditional ways and for passengers' comfort during flight, surveys have demonstrated passengers' preference for connectivity on-board other than food and with allowance for personal electronic devices it is difficult to reverse this trend.

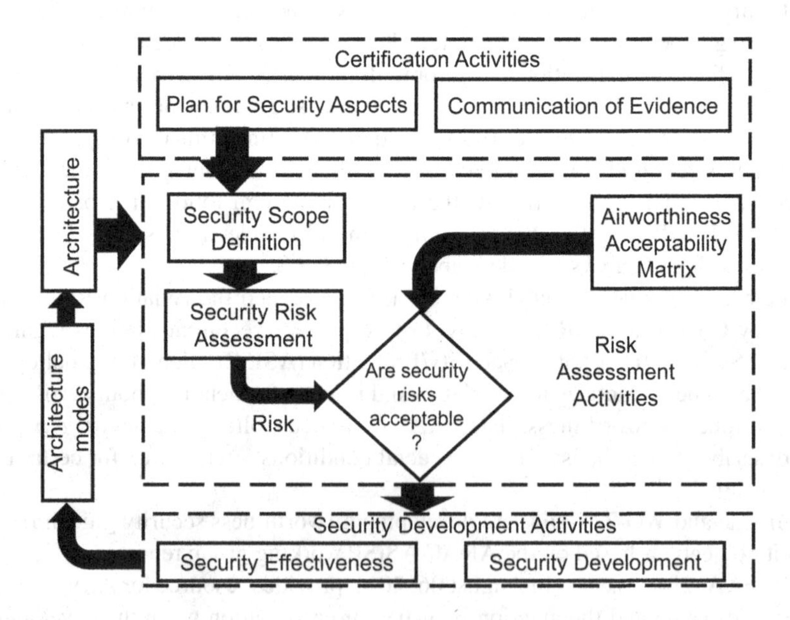

FIGURE 12.1 Certification process.

To respond to these needs airliners and manufacturers are considering connectivity from early design stages throughout the life cycle. It is important that cybersecurity be part of the system from the beginning and not added on later.

Most of in-service aircraft are in an operational life-cycle phase, which make them eligible to modifications through service bulletins and supplemental type certificates that can introduce new functionalities which in turn can transform a prior "isolated" aircraft in a "connected" aircraft and all cybersecurity risks that this modification will introduce.

According to Falco et al. (2019), "it may be possible in certain domains [e.g.; aviation] to eliminate certain kinds of cyber risk." They also state that "managerial options to avoid cyber risk include minimizing the use of connected computing systems in certain environments."

Beyond on-board connectivity, this section presents the aircraft embedded systems architectural aspects, which aims to reduce weight, maintenance, and improve resources. Solutions like field loadable software and shared data buses increase architectural complexity. Furthermore, a lack of proper analysis will introduce vulnerabilities that are not identified and mitigated as described in FAA (1988).

REFERENCES

EASA. 2016. Aircraft Cybersecurity European Aviation Safety Agency.

Eurocae, and WG-72. 2014. *Airworthiness Security Process*. Washington, DC: FAA.

FAA. 1988. Advisory Circular 25.1309-1A—System Design and Analysis. Washington, DC: Federal Aviation Agency.

FAA. 2014. *Establishment of Special Conditions for Cybersecurity*. Washington, DC: Federal Aviation Agency.

FAA. 2019a. "Function and Installation (14 CFR 25.1301)." Federal Aviation Agency, accessed 11 August. https://www.law.cornell.edu/cfr/text/14/25.1301.

FAA. 2019b. "Regulatory and Guidance Library." Federal Aviation Administration, accessed 11 August. http://rgl.faa.gov/

Falco, Gregory, et al. 2019. "Cyber Risk Research Impeded by Disciplinary Barriers." *Science* 366(6469):1066–1069.

RTCA. 2014a. *Airworthiness Security Methods and Considerations*. Washington, DC: RTCA.

RTCA. 2014b. *Information Security Guidance for Continued Airworthiness*. Radio Technical Commission for Aeronautics.

13 Safety

There is a strong relation between safety and the systems approach. It is not that safety is part of the systems approach or *vice versa.* However, many systems engineering principles are used in safety analysis, such as requirement flow-down. The same is true of cybersecurity analysis.

As shown in Figure 13.1, fatality rates documented by Gilligan (2005) have fallen over the years. This chart shows how the different technology advances contributed to this drop. Another possible contributor to this drop is the basic systems engineering principle of verification. That is to say, as was shown in Chapter 5, if a component is verified to the worst-case conditions, and if conditions never worse than that condition are encountered, no component failures should be expected.

This section does not attempt to explain all facets of safety analysis, but rather to show the relation between safety and systems principles. Figure13.2 explains that relation. The general principles of system safety for commercial aircraft can be found in FAA (2000).

To understand this chart, the reader should keep in mind the concept of hierarchy described in Chapter 6. The systems engineering items on the left-hand side of the chart pertain to higher levels of the aircraft physical hierarchy. For example, the requirements management and requirements items on the left pertain to the requirements being flowed down from the next higher level of the aircraft hierarchy. For example, the requirements at that level may be subsystem level requirements.

These subsystem requirements then enter the safety domain where new requirements will result. A common safety requirement is spacing, that is, it pertains to how far apart individual components need to be spaced. The new spacing requirement then passes to the right side of the chart that pertains to hierarchy level of interest. Obviously that requirement will need to be validated, which is done on the right side of the chart.

The right side of the chart also introduces other considerations. For example, the spacing requirement may introduce new risks that need to be evaluated here. Obviously, the configuration of the subsystem will change; which will need to be recorded in the configuration task. In addition, the introduction of a new spacing requirement may introduce a new interface; this aspect will also need to be treated on at this time and level.

The above example of a safety requirement is just one of many possible Safety requirements that may be introduced and stimulate an interaction between Safety and Systems Engineering.

SAFETY AS AN EMERGENT PROPERTY OF SYSTEMS

This section shows the idea of systems thinking that safety is an emergent property of the system and that all the consideration of systems theory and the systems approach are a strong foundation to address the safety aspects that need to be considered in

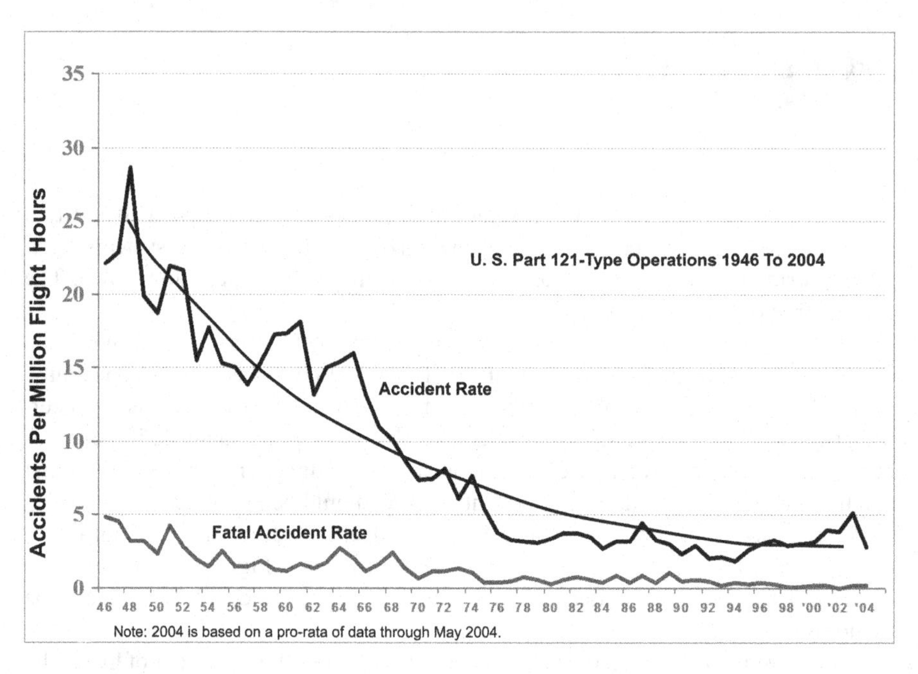

FIGURE 13.1 Fatalities.

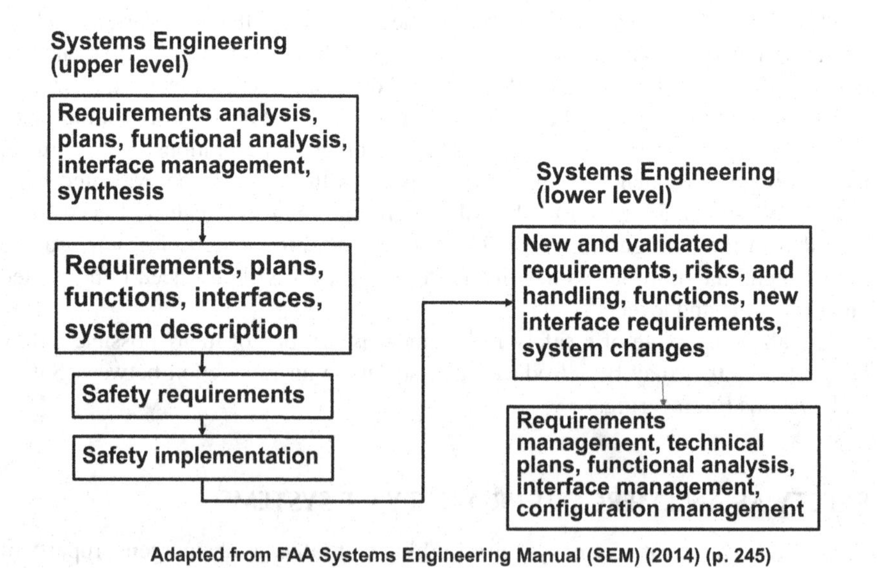

FIGURE 13.2 Safety and systems engineering.

the early phases of solution definition. These aspects involve the architecting process on the idea of conduct architecting activities with safety and security considerations on the concept level of the system. STAMP (System-Theoretic Accident Model and Processes) created by Professor Nancy Leveson at MIT, is a good technique to help the architects address system level safety analysis and security scenarios. STAMP is the name of the new accident causality model based on systems theory. In STAMP, safety is treated as a dynamic control problem rather than a failure prevention problem.

SAFETY AND DECISIONS

The second concept, which has a strong relevance to human error, is the concept of decisions. Decisions are a two-sided sword. On one side there are the decisions dominated by cognitive bias, which is the tendency to make decision errors as a result of prior beliefs, emotion, and/or lack of information. Cognitive bias has only recently become recognized as an important contributor to accidents; however, the Commercial Aviation Safety Team (CAST 2011) recognizes it as a factor. In addition, the FAA has accumulated a collection of common themes judged to have contributed to accidents. These themes documented in FAA (2019) include flawed assumptions, human error, organizational lapses, pre-existing failures, and unintended effects. Many of these themes imply the cognitive biases discussed in this chapter. For example, the definition of cognitive bias includes the presence of "prior beliefs" that are part of the flawed assumptions described by the FAA.

The other side of the sword is the effect of beneficial emotions. So emotions can have both positive and negative effects. According to Damasio (1994, p. 53), "Reduction in emotion may constitute an equally important source of irrational behavior." According to Lerner et al. (2015), some emotions, such as happiness, pride, and love may often, but not always lead to enhanced decisions. Another characteristic known as "high emotional intelligence" may also be beneficial to decisions.

This section first discusses irrationality especially as it pertains to cognitive bias. It also discusses case studies in which cognitive bias is thought to have played a role. These cases are divided into three categories that reflect the degree of confidence in which cognitive bias is thought to have played a role: possible, probable, and firm. This confidence is the result of analyses by the authors based on the evidence in the case studies and on the analysis of other authorities. The two types of biases discussed are individual and organizational. Each case study will describe the consequences that resulted from the cognitive biases. Finally it discusses several mitigation approaches that that experts have suggested.

The importance of cognitive bias was brought to light by the awarding of the Nobel prize to both Daniel Kahneman and Richard Thaler for their work in this subject. Their findings are documented in the books by Kahneman (2011) and Thaler and Sunstein (2008). In short, Kahneman and Thaler concluded that people make mistakes in decisions due to these factors: prior beliefs, emotion, and/or lack of information.

They conclude that most, if not all, people suffer from cognitive bias. More amazingly they conclude that intelligence, importance of the decision, and time to think do not improve the decisions, at least not much. Thaler goes one step further and suggests that there are many ways to influence decision makers to make better decisions. However, these influences are all *external*, not internal.

Another contributor to this field is Reason (1997), who concluded that organizational characteristics were an important factor in accidents. Reason introduced the concept of organizational accidents.

Jackson and Harel (2017) expand on this topic and show that organizational characteristics can combine to cause catastrophic accidents. Furthermore, Jackson and Harel show how this topic fits into systems engineering context. The reason is, in short, that decision management is an important process within systems engineering, and anything that would contribute to errors in decisions is important. Furthermore, anything that would influence better decisions is also good.

Traditional literature, for example, International Council on Systems Engineering (INCOSE) (2015), on decision-making especially within systems engineering has assumed that decision makers can make rational decisions. Kahneman and Thaler have shown that this assumption is highly flawed.

Murata, Nakamura, and Karwowski (2015) also contributed to this field by examining catastrophic accidents showing how cognitive bias was a factor. Other researchers, for example, Kahneman and Thaler, focus primarily on low consequence decisions, for example, whether to go on a diet or not. Murata et al. show that the same principles can apply to high-consequence decisions.

A key concept in the study of cognitive bias is *irrationality* which is the predominance of prior beliefs and emotion over goals and information. The regions of rationality and irrationality are shown in Figure 13.3. However, this chart pertains only to those emotions for which decisions are degraded. It will be remembered according to Damasio, above, that a reduction in emotion can often result in an improvement in rationality.

Cognitive bias can be divided into two categories: individual bias and organizational bias. Individual bias is that bias experienced by an individual and not influenced by its presence in an organization. Even though many people in an organization

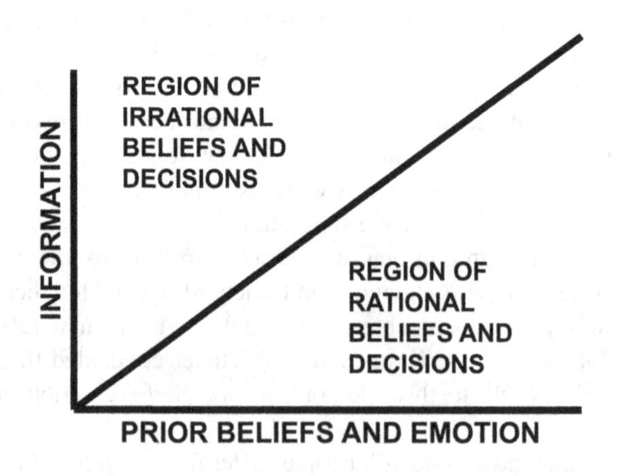

FIGURE 13.3 Beliefs and goals.

suffer from individual bias, it is not considered an organizational bias unless some characteristic of the organization influences that bias. It can be concluded then that organizational bias is a bias influenced by a characteristic of the organization. Three characteristics have been identified: the hierarchical structure, culture, many interacting people.

Some typical individual biases are as follows:

- **Bias denial.** The denial that biases exist or than decision maker suffers from them.
- **Confirmation bias.** The view evidence that prior belief is true even if is not.
- **Complacency bias.** The situation in which the decision maker is complacent.
- **Optimism bias.** Belief that system will work.
- **Normality bias.** Belief that perturbations are normal and will not have serious consequences.
- **Risk denial.** Belief that there are no risks (high priority).
- **Plan continuation.** The continuation of tasks in spite of warnings.
- **Visual assessment.** Overconfidence in assessment of situation.

Some typical organization biases are as follows:

- **Authority bias.** Trusting too readily in a person of authority.
- **Rankism.** Assumption that person of higher rank has superior decision capability.
- **Culture bias.** Inattention to risk and safety by members of an organization.
- **Groupthink.** Overestimation of the group, pressure toward conformity.
- **Protocol rationalization.** Assumption that protocol has higher priority than common sense.
- **Responsibility bias.** Exaggeration of one's own contributions.
- **Accountability bias.** The refusal to take responsibility for problems.
- **Loss aversion.** The fear of making a wrong decision even if it is right.
- **Social loafing.** The tendency to exert less effort on the assumption that someone else is doing the job.

ORGANIZATIONAL BIASES

In addition to the biases suffered by individuals described above, there is another category of biases. These are called *organizational biases*. Individual biases are those suffered by individuals whether or not they are part of an organization. Organizational biases are those biases suffered by individual who are part of an organization. These biases are caused by specific features of an organization. These features are as follows:

- **Hierarchical structure.** Almost all organizations are structured hierarchically. There is the entire organization at the top of the hierarchy, then followed by divisions and departments below.

The organizational hierarchy gives risk to the *rankism* bias; that is the belief that one's position in the hierarchy gives that person not only authority but also superior knowledge as to what the right decision might be. This bias has been found to be particularly prevalent in the cockpits of airplanes where the captain assumes a superior position and refuses to listen to the warnings of other crew members.

The *authority* bias is also influenced by the hierarchical structure.

- **Many interacting human beings.** This feature of organizations tends to amplify such biases as *complacency.* This factor tends to exacerbate the lack of communication within an organization. Complacency is the inattention to risk and safety. *Groupthink* is another bias influenced by this factor.
- **Culture.** Many organizations have a culture of *complacency*, for example. Although individuals may suffer from complacency, often this bias exists on an organizational level.

EXAMPLE ORGANIZATIONAL BIASES

The organizational biases listed below are only a select group from the hundreds in the literature that may have been contributing factors in catastrophic accidents. It must be emphasized that organizational biases do not alone lead to accidents. For example, almost all organizations are structured hierarchically. Yet this characteristic alone does not always lead to accidents. The human is an essential ingredient in accidents.

Authority bias. This is an organizational bias because it requires the existence of multiple layers of authority. It is closely related to the *rankism* bias. This bias may have been a factor in such accidents as the Tenerife accident, the Korea 801, or the Air Blue 202 accident when the aircraft crew deferred to the captain in a moment of crisis.

Rankism bias. This bias turned out to be an important factor in several accidents including the Tenerife accident, the Korea 801, and the Air Blue 202 accident. In all of these cases the captain assumed the authority to make decisions simply based on his rank, which is the core aspect of *rankism.* This bias is founded in the hierarchical characteristic of organizations.

Culture bias. This is the collective disregard for safety and risk. It is shared by many people; therefore, it is an organizational bias. Culture has been found to be a major factor in many catastrophic events, for example, the *Challenger* disaster, as documented by Vaughn (1997). Within this culture Vaughn found that risk was considered to be "normative."

Groupthink bias. This is the overestimation of the group and the pressure towards conformity. This bias qualifies as an organizational bias since many people are involved, not necessarily many people of a higher rank, as in the *rankism* bias, just many people. It has been speculated that the *groupthink* bias was at work in the Honda Point disaster in which several ships were grounded simultaneously.

Protocol rationalization bias. This is the assumption that protocol has higher priority than common sense. This bias has not been found in any literature on cognitive bias. However, it was derived from the United 93 case study in which the FAA

(Federal Aviation Agency) refused to warn the pilot of this plane that there were terrorists on board because it was a violation of their protocol.

Responsibility bias. This bias is the exaggeration of one's own contributions.

Accountability bias. This bias is the refusal to take responsibility for problems. This bias was seen to be a factor in the Deepwater Horizon event.

Loss aversion bias. This bias is the fear of making a wrong decision even if it right. This bias can be a factor when the decision maker is trying to meet the expectations of higher-level persons.

Social loafing bias. This bias is the tendency to exert less effort on the assumption that someone else is doing the job.

MITIGATION APPROACHES

This section summarizes the approaches to minimizing the effects of cognitive bias. Mitigation includes *debiasing*, that is, the removal or reduction of the bias itself and also the reduction of the *effects* of cognitive bias. These approaches are taken from several sources.

SELF-MITIGATION

Self-mitigation is the mitigation of a bias by the decision maker through training or simply logic. Kahneman and Thaler cast doubt whether this can be done for all biases and by all people. They make three points, that bias mitigation is independent of the:

- Intelligence of the decision maker.
- Importance of the decision.
- Time spent thinking about the decision.

Nevertheless, it is fair to say that some biases are easier to mitigate than others.

It is safe to say that biases can be self-mitigated, that is by common sense, if (a) the risk of the bias is low and (b) the stress that led to the bias is also low. Chapter 4 discusses some of the stresses and pressures can lead to the need for more advanced methods of mitigation. Figure 13.4 shows this region of self-mitigation. It also shows that if the stress is high or the risk is high, more advanced methods of mitigation are required. This section summarizes some of those methods.

AUTOMATED MITIGATION

In some cases the mitigation approach is so obvious it can be implemented automatically without the intervention of a human. A case in point is traffic collision avoidance system (TCAS). This system that is on most commercial aircraft acts to prevent a mid-air collision between two aircraft on a predicted collision path. This system is normally on both aircraft. It causes one aircraft to maneuver up and the other aircraft to maneuver down, thus avoiding each other. There is no stress or other pressures in this situation or the actions to avoid a collision. It happens automatically.

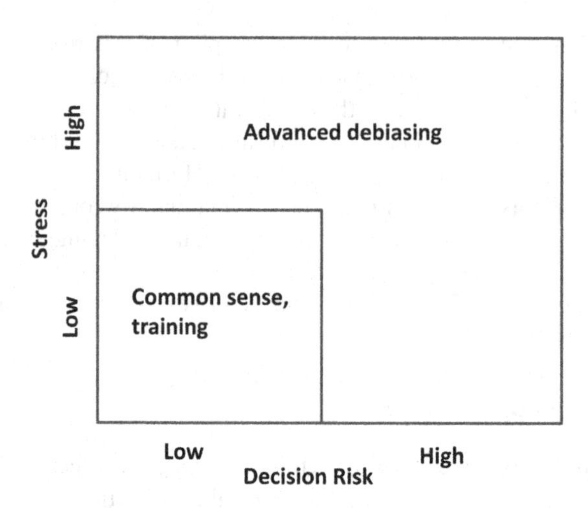

FIGURE 13.4 Regions of mitigation.

GROUP METHODS

The premise of all group methods is that the wisdom of a group is better than an individual (except in the case of the *groupthink* bias discussed above). The limitation of all group methods is that they involve a time delay that may be too long to prevent an accident.

The premortem. According to Kahneman (2011, pp. 264–265), the technique of the premortem is to gather a group of trusted advisors to advise on decisions. The job of these advisors is to focus on possible negative consequences. Kahneman states that the premortem has two main advantages: (1) it overcomes the group think that affects many teams once a decision has been made and (2) it unleashes the imagination of knowledgeable individuals in a much-needed direction. He goes on to say that it "goes a long way toward reducing the damage of plans that are subject to the biases of WYSIATI (what you see is all there is)."

Communities of practice. This is a method described by Jackson (2010, Chapter 6). The basic aspect of this method is self-learning between management and employees. There is a circular process of information and learning. This method depends on management's willingness to accept lessons from employees and *vice versa*.

Crew resource management (CRM). This is a practice that has been accepted within the aviation domain. The basic purpose of it is to train crew members who know and understand their responsibilities. One of those responsibilities is to inform the captain of any impending risks, such as flying into a mountain. This method has been found to be effective in countering the *rankism* bias.

Consensus. The idea behind a consensus is that if a small group of people agree on an issue, they are probably right. According to Thaler and Sunstein (2008, pp. 57–58), people in small groups usually come to an agreement very quickly. The limitation of this method is that a different group answering the same question may arrive at a different answer. This is not to say that the consensus answer is wrong; it just says that there may be other ways to arrive at an answer.

Pre-commitment. The idea behind a pre-commitment is that decisions can be made in advance before the issue even arises. One of the preceding techniques can be used to make the decision. The limitation is that the circumstances behind the decision need to be known. No surprises are allowed. So to the extent to which the circumstances are known, this technique may be valuable.

Policy change. It has been found that policies in certain organizations contributed to the stress that led to one or more accidents. As an example, it has been concluded that the pilot rest policy at an airline contributed to the "hurry-up" stress that led to the Tenerife accident described in Chapter 3. A simple remedy to this problem is to eliminate or modify this policy. It is assumed that this was done after that accident.

Protocol change. Similar to policy change, elimination or modification of protocols that have led to accidents may lessen the chance that these types of accidents will occur. As an example it was seen that the *protocol rationalization* bias contributed to the United 93 accident described in Chapter 3.

Independent review. This is one of the most highly regarded of all mitigation methods. For example, it is the only method endorsed by the Columbia Accident Investigation Board. The concept is that an independent board would be available to advise decision makers before any high-risk decisions.

Independent means that the board would be both organizationally and financially independent of the program in question. These rules protect against conflicts of interest. That is to say, the board would need to be part of another program or another company and not part of the organization being reviewed.

SUMMARY OF COGNITIVE BIAS CONCLUSIONS

Based on the literature of well-known accidents, it is clear that more fatalities have resulted from cognitive bias than is generally recognized. This is particularly true in the aviation domain where organizational factors contribute to cognitive bias.

Furthermore, most, if not all humans, experience cognitive bias in their daily life. As was shown in this section and verified by respected psychologists, in most cases intelligence is not a barrier to cognitive bias. Individuals can only mitigate cognitive bias if the stress is low and the risk is low. Experts have suggested mitigation approaches for these biases. Some of them have been shown to be effective and have been accepted into practice.

As a recommended action, OEMs should review nudge methods to reduce decision errors in design, maintenance, and repair. Independent review is recommended. These methods should be incorporated into standard procedures.

SUMMARY OF EMOTION AND COGNITIVE BIAS ON FLAWED DECISIONS AND SAFETY

From the above it has been concluded by authorities that decisions can be either enhanced or degraded by factors such as cognitive bias and emotions, both positive and negative. It is the degraded decisions we are most concerned about from a safety point of view.

Figure 13.4 gives an overview of both enhanced and degraded decisions caused by these factors.

This figure merges ideas from Kahneman (2011), Thaler and Sunstein (2008), and Lerner et al. (2015). Kahneman, Thaler, and Sunstein focus on cognitive biases and their tendencies to degrade decision capability. Lerner et al. focus on the many aspects of emotion to drive decision capability, especially the capability to enhance decisions. The central idea of the figure is that emotions, stresses, prior beliefs, and emotional intelligence can go either way. Under certain conditions, emotions can degrade decisions; under other conditions emotions can enhance decisions. The literature on emotions does not give many examples of which emotions enhance decisions and which ones inhibit decisions. For example, the emotion of love will probably enhance decisions, while anger could have the opposite effect. Thus, in the end it is up to the analyst to take steps to enhance decisions or to mitigate their degradation. In the end whether emotions degrade or enhance decisions will largely depend on the individual emotions and the expected outcome.

The first block in Figure 13.5 shows some of the emotions mentioned by Lerner et al. (2015) that may either enhance or degrade decision capability. The stresses and pressures mentioned in the block below are more than likely to degrade decision capability. Then prior beliefs, such as that the pilot is more capable of making decisions than the ATC, will most likely degrade the decision capability. Finally, as the Lerner et al. (2015) state, some people have superior emotional intelligence that allows them to make better decisions regardless of their emotional state. When all

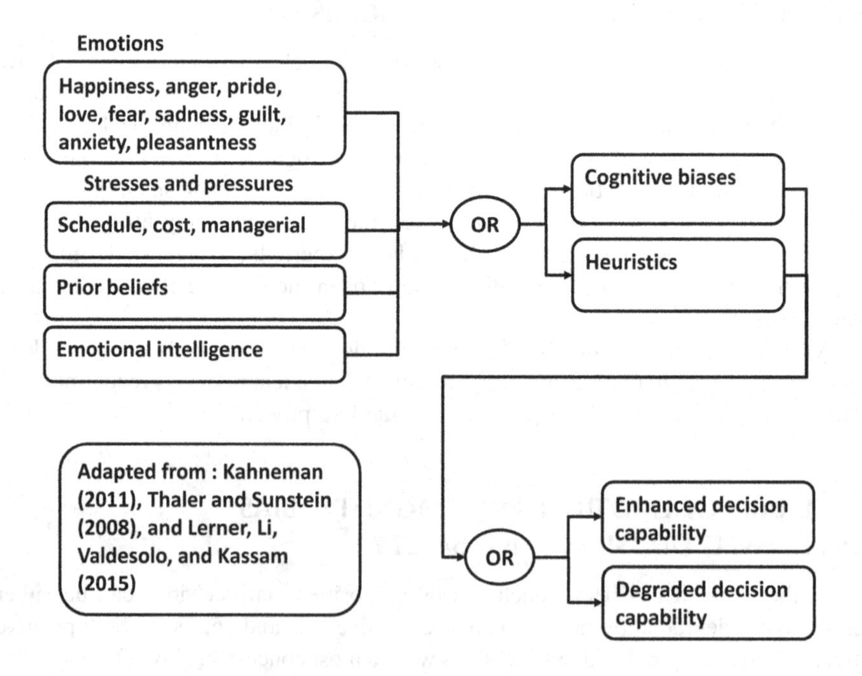

FIGURE 13.5 Decision path.

of these factors are combined, one's decision capability may be either enhanced or degraded according to which influences were strongest.

So after these influences are combined, they are reflected in either cognitive biases (usually the negative impacts on decisions) or heuristics (usually the positive impacts on decisions).

SAFETY AND ACCIDENTS

The accidents discussed in this section are a selection of incidents in which loss of life occurred. It is therefore a part of safety in the broadest sense. In particular this section will discuss incidents in the context of systems engineering and the SE principles that were apparently ignored or conducted ineffectively. This focus underscores the strong relation between SE and Safety.

The accident causes in this chapter focus on two aspects: flawed decisions due to cognitive bias as discussed above and also the failure to verify requirements to a worst-case scenario which is an SE fundamental. Flawed decisions can occur at any time during the aircraft life cycle and can therefore contribute to accidents. These decisions include design, integration, production, test, maintenance, and operation.

Therefore, in the end, the focus of this section will be how to mitigate the cognitive bias and to assure that all components are tested to worst-case conditions. Neither of these is as easy as it may seem, so the following paragraphs will elaborate on each one.

DECISION ERRORS BY LIFE CYCLE PHASES

Decision errors resulting in catastrophic consequences may occur at any phase of the aircraft life cycle. We will focus on four phases: the design phase, the operations phase, the maintenance phase, and the repair phase. Catastrophic decisions have been made in all these four phases.

The design phase. During the design phase, the design organization makes key decisions: the requirements for the design, selection of the design features, and verification that the design meets the requirements. The latter is the most critical because if the design does not meet the requirements under the most stressful conditions, the likelihood of failure will be high.

The operations phase. The success of this phase rests largely, but not completely on the pilot. There are numerous cases in which pilots suffering from cognitive bias caused catastrophic accidents, as will be seen later.

Maintenance phase. Catastrophic accidents resulting from maintenance errors are rare, but they do happen especially when the maintenance is not performed in accordance with prescribed procedures. American Airlines Flight 191 was such a case as documented by NTSB (1979).

Repair phase. Similarly, there are very few accidents due to faulty repair. Japan Air Lines 123 described by Japan Air Lines (1985) was such a case. In this case the aircraft was damaged following a tail strike and then repaired. It raises the question of how well repairs should be executed.

Recommended practice. The recommended practice (SAE 2010) developed for the FAA is consistent with systems principles especially with respect to the development and verification of requirements. Specifically it states, "The verification of implementation against allocated requirements may be performed at the system level or at the item level." The failure to comply with this statement has contributed to several catastrophic events as this chapter shows. The requirement for an item (a component) to be verified implies that the failure to verify the requirement may result in a catastrophic accident. If the requirements for a whole system (a subsystem) to be verified imply that all the components in that system also comply with requirements. This document ARP 4754A also states that "worst case issues" should be covered. Worst-case issues refer to the worst-case environments in which the item or system has to perform.

When a component fails a worst-case condition, there are two possible reasons: The first is that component was not verified to that condition. This can happen and has happened. The other reason is that the actual conditions exceeded the design limits. This happens frequently with respect to bird strike and occasionally with respect to lightning. The conclusion is that all components need to be verified to worst-case conditions including repairs and modifications. This latter verification is not always performed and there have been catastrophic events as a result of this lack of verification.

The Tenerife accident. Let's start with the worst accident in commercial aviation history which is generally known as the Tenerife accident. This accident is described by Brafman and Brafman (1998) and also McCreary et al. (1998). Furthermore, FAA (2019) describes this accident in which the pilot was in a hurry to take off due to crew duty time limitations putting this in the *bias denial* or *risk denial* category. Brafman and Brafman (1998) agree with the FAA and go to great lengths to explain in psychological terms why a pilot would be willing to make such a decision with such a serious consequence. McCreary et al. (1998) examine the same situation and conclude that the stress of the situation limited the ability of the pilot to consider long-term consequences. Another possible bias is the *rankism* bias in this case, which says that your decision is supported by your higher rank in the organization. As a result of these analyses, this accident belongs in the *probable* category for cognitive bias consideration.

Air Blue Flight 202. Another good example of cognitive bias is Air Blue Flight 202 from Karachi to Islamabad which in 2010 the Airbus 321 flew into a mountain in Pakistan killing all on board. This is a perfect example of the *rankism* bias. According to Pakistan CAA (2011), the pilot began the flight by warning the first officer not to say anything and to be quiet. Later in the flight when the ATC ordered the pilot to turn, he would ignore the ATA and perform a different maneuver. These are two examples of why this case study qualifies as a *firm* example of cognitive bias.

Alaska Flight 261. In the year 2000 an Alaska flight 261 MD-83 crashed off the coast of California when the jack screw in the tail failed due to improper maintenance. There were 88 fatalities. According to NTSB (2002), the cause of the accident was the lack of lubrication in the tail assembly jack screw and the extended intervals of maintenance. From these facts and the BTSB findings, the probable root cause of the accident was the cognitive bias of *complacency*.

Asiana Flight 214. In 1914 a Boeing 777 collided with a wall at the end of the runway in San Francisco. There were three fatalities and 301 survivors. The root causes of the accident were judged to be *complacency* and lack of experience of the flight crew. According to NTSB (2014), contributing to the accident were the flight crew's "non-standard communication." As a result of the above analysis, a cognitive bias of *complacency* is assigned as a *possible* factor in the accident.

Japan Air Lines 123. In 1985, a Boeing 747 crashed on August 12 having suffered a tail strike seven years earlier. According to Japan Air Lines (1987), the repairs to the aircraft appear to have been "proper." However, in mid-flight the control of the aircraft became impossible. In the end, the aircraft crashed near Mount Takamagahara with 520 fatalities and four injuries. The repairs to the aircraft did not conform to Boeing's own approved repair process.

The lessons from this accident are, first that repairs to damaged aircraft should be conducted in accordance with approved methods. The highest level of approval would be the regulatory agency. These methods are normally documented in a company repaired SRM (structural repair manual).

Second, the repair should conform to the quality criteria specified in the SRM and approved by the regulatory agency. For example, should the repaired structure be as strong as the structure before the damage occurred. This would be the ideal and most logical criterion.

Finally, who would approve the quality of the repaired structure? Ideally this would be a regulatory agency function or a representative of the regulatory agency.

In short, all work should be performed by certified persons, inspected by certified persons, and meet regulatory requirements. The failure to do any of these things may result in the catastrophe such as the JAL 123.

AA Flight 965, the Calí Accident. According to the Colombian Civil Aeronautical Authority (1996), this aircraft was approaching the Calí airport in Colombia when it was directed to change runways. The pilot pressed a button which had the effect of incorrectly directing the aircraft to another airport. This type of bias is known as a *slip* bias that occurs when there is a lack of time or a confusing cockpit. The result was that the aircraft flew into a mountain killing 159 people and injuring four.

The report recommends a number of remedies some of which included changes to the FMS (flight management system). The slip bias can be regarded as a *probable* cause.

O'Hare AA Flt 191. According to NTSB (1979), in 1979 the carrier American Airlines had decided to perform maintenance of the engines on the tarmac rather than in the hangar as was required on the DC-10 at the Dallas-Fort Worth airport. The change in maintenance procedure resulted in added stress on the engine bolts that resulted in the loss of one engine upon takeoff. The aircraft spiraled out of control and crashed. All 272 persons on board perished.

The root cause of this accident was a maintenance error resulting from *complacency* and *social loafing*. Based on the evidence at hand these biases are considered *probable* causes of this accident. Additional factors included the time and cost of maintaining the aircraft in the conventional way.

United 93. This is a well-publicized incident that was part of the 9/11 event described by the 9/11 Commission (2004). In this case the aircraft was delayed taking off, so it was airborne when the attacks on the World Trade Center occurred.

The 9/11 Commission strongly criticized the FAA for failing to notify the pilots directly when it was known that there was a possibility of terrorists on board. The FAA responded that it was not part of standard protocol to notify pilots directly but to notify the airline. This protocol resulted in a delay that resulted in the crash of the aircraft.

The 9/11 Commission stated that the FAA did not understand their responsibility. We do not claim that any individual or organization was responsible for the fatalities that ensued, other than the terrorists. Our sole purpose is to identify the cognitive bias in this case. We call it the *protocol rationalization* bias. This bias states that protocol takes precedence of any other consideration including rational thought.

It is not known whether the FAA has changed their internal policies regarding communication or not. Such information is not publicly disclosable.

United 585 and USAir 427. In the years 1991 and 1994, two 737s crashed in Colorado Springs and Pittsburgh for the same reason, namely, the failure of a rudder actuator. There were 25 and 132 fatalities, respectively. The accidents are summarized in NTSB (2001) and NTSB (1994). Since a rudder actuator is a component of an aircraft that is critical to its safety, it would be expected that this actuator had been tested in all anticipated environments since this is a basic systems engineering principle. It appears, therefore, that this testing was not done perhaps because this was not an anticipated environment. Nevertheless, these failures show that this is still a valid rule and if these actuators had been tested, the likelihood of failure would have been less likely. A *possible* cognitive bias for these accidents is *complacency.*

Korean 801. This is a landmark accident especially with respect to the *rankism* bias, which qualifies as a *firm* type of bias. According to NTSB (2000), a 747 impacted a hill on its approach to Guam in 1997. There were 228 fatalities. In this accident the captain failed to heed the warnings of other crew members who were trying to warn him of the impending crash. Another factor is that one instrument was out of service. This is one of several accidents that led to the CRM policy.

Air France 447. This is another accident that emphasizes the importance of component testing. According to BEA (2012), this aircraft, an Airbus A330, lost control in heavy weather off the coast of Brazil in 2009. This loss of control was mainly due to the failure of the pitot tubes due to icing. The report also cites the inexperience of the pilot in this situation.

Once again it raises the question of whether the pitot tubes had been tested in this condition and if not, why not? A *possible* cognitive bias was therefore the *complacency* bias.

Nagoya. In 1994 an Airbus A300 was attempting to land at the Nagoya airport in Japan. According to Ladkin (1996), the pilot inadvertently put the aircraft in the go-around mode through a *slip* error. In short, the pilot wanted to land but the aircraft wanted to go around. The net result was that the pilot and the aircraft were in conflict and the aircraft stalled and crashed with 264 fatalities.

Another factor not mentioned in the report was the failure of the design to implement the Billings rule (Billings 1997) that every component of the system should understand the intent of every other component. In this case, one component was the pilot and the other component was the flight control system. In this case the firm cognitive bias was the *slip* bias.

Turkish Airline. According to Dutch Safety Board (2010), the immediate cause of this crash of the 737 was the failure of the radio altimeter. There were 9 fatalities. From a cognitive bias point of view, the *probable* cause was the *complacency* bias that caused the airline not to train its pilots how to handle a failed altimeter.

Summary by phase. From the selected case studies summarized above, it is seen that cognitive biases in almost any life cycle phase can contribute to accidents. Following is a summary of accidents by phase according to life cycle:

- **Operational phase.** Air Blue 202, Asiana 214, Calí, United 93, Korean 801, Nagoya
- **Design phase.** United 427, United 585, Air France 447, Nagoya
- **Maintenance phase.** Alaska 261, AA 191
- **Repair phase.** JAL 123

COMPONENT FAILURES

As mentioned at the beginning of this chapter, the focus will be on two types of failures: failures due to irrational decisions (cognitive biases) and component failures. In recent years component failures have been rare, and the number contributing to catastrophic failures has been even rarer. The basis causes of component failure are as follows: failure to verify the component requirement to the worst-case condition, faulty maintenance, and encountered conditions which were beyond reasonably expected levels.

United 585 and USAir 427. These cases were described above when the rudder actuator failed when they attempted to execute actions beyond the design limits. It can only be assumed that these actions were beyond the normal design boundaries, so that verifying the requirements for them would not have been possible. Nevertheless, this fact does not diminish the importance of the principle of verifying all requirements to the worst-case conditions. In this case, this was not done. In the future, if possible, this should be done.

Alaska Flt 261. The component in this case was the jackscrew in the tail assembly. It was determined that it failed because of inadequate maintenance and the use of an improper lubricant. In both cases the *complacency* bias comes to mind. Hence, component failures can also involve cognitive biases.

All Nippon battery fire. In this case the cause of the battery fire has been described as excessive complexity. In any case it is clear that the battery in question did not perform to expected levels. The most obvious bias in this case is the *optimism* bias. The failure to verify this battery to its expected performance can be laid to the *complacency* bias.

Delta electric generator. An electric generator was installed on a Delta MD-90 causing a large amount of EMI (electromagnetic interference) and the inability of the avionics system to work. No accident occurred, but expense was incurred fixing the problem. The root cause of the problem was the failure to provide a procurement specification for the battery with EMI requirements in it. This can be regarded as a complacency bias example.

Turkish Airlines Flight 1951. In 2009 a Turkish Airline 737 crashed on approach to Amsterdam. There were multiple fatalities. The root cause of the crash was a faulty altimeter, which was the component in question. It can be assumed that the requirements for this components had not been verified. Furthermore, the fact that there did not seem to be a redundant altimeter was apparent. In any event the active bias was the *complacency* bias.

REFERENCES

9/11 Commission. 2004. 9/11 Commission Report. edited by Thomas H. Kean. Washington.

BEA. 2012. Air France 447 Final Report. Paris: Bureau d'Enquêtes et d'Analyses pour la sécurité de l'aviation civile.

Billings, Charles. 1997. *Aviation Automation: The Search for Human-Centered Approach.* Mahwah, NJ: Lawrence Erlbaum Associates.

Brafman, O., and R Brafman. 1998. "Anatomy of Disaster [Tenerife]." In *In Sway: The Irrisistable Pull of Irrational Behavior*, 9–24. New York: Crown Business.

CAST. 2011. "The Commercial Aviation Safety Team." Last Modified 2011, accessed 16 February. http://www.cast-safety.org/about_vmg.cfm.

Colombian Civil Aeronautical Authority. 1996. AA965 Cali Accident Report, edited by Peter Ladkin. Bogota, Colombia: Aeronautica Civil of the Republic of Colombia.

Damasio, Antonio. 1994. *Descartes' Error: Emotion, Reason, and the Human Brain.* New York: Penguin Books.

Dutch Safety Board. 2010. Crashed During Approach, Boeing 737-800, near Amsterdam Schiphol Airport, 25 February 2009. The Hague: Dutch Safety Board.

FAA. 2000. *FAA System Safety Handbook*, edited by Safety. Washington DC: Federal Aviation Administration.

FAA. 2019. "Accident Common Themes." Federal Aviation Authority, accessed 19 May. https://lessonslearned.faa.gov/ll_main.cfm?TabID=4.

Gilligan, Margaret. 2005. "Title." European Civil Aviation Conference.

International Council on Systems Engineering (INCOSE). 2015. *INCOSE Systems Engineering Handbook: A Guide for System Life Cycle Processes and Activities.* Fourth ed. Hoboken, NJ: ICOSE.

Jackson, Scott. 2010. "Architecting Resilient Systems: Accident Avoidance and Survival and Recovery from Disruptions." In Wiley Series in Systems Engineering and Management, edited by Andrew P. Sage. Hoboken, NJ, USA: John Wiley & Sons.

Jackson, Scott, and Avi Harel. 2017. "Systems Engineering Decision Analysis can Benefit from Added Consideration of Cognitive Sciences." *SyEN* 55, 19 July.

Japan Air Lines. 1985. Aircraft Accident Investigation Report (JAL 123). Gunma Prefecture: Japan Air Lines.

Japan Air Lines. 1987. Aircraft Accident Investigation Report (JAL123). Gunma Prefecture: Ministry of Transport.

Kahneman, Daniel. 2011. *Thinking Fast and Slow.* New York: Farrar, Straus, and Giroux.

Ladkin, Peter B. 1996. "Resume of the Final Report of the Aircraft Accident Investigation Committee into the 26 April 1994 Crash of a China Air A300B4-622R at Nagoya Airport, Japan." In *The Nagoya A300-600 crash.* Bielefeld, UK: University of Bielefeld—Faculty of Technology.

Lerner, Jennifer, Ye Li, Piercarlo Valdesolo, and Karin Kassam. 2015. "Emotions and Decision Making." *Annual Review of Psychology* 66:799–823.

McCreary, John, Michael Pollard, Kenneth Stevenson, and Mark B Wilson. 1998. "Human Factors: Tenerife Revisited." *Journal of Air Transportation World Wide* 3(1).

Murata, Atsuo, Tomoko Nakamura, and Waldemar Karwowski. 2015. "Influences of Cognitive Biases in Distorting Decision Making and Leading to Critical Unfavorable Incidents." *Safety* 1:44–58.

NTSB. 1979. American Airlines Flight 191 Accident Report. Washington: National Transportation Safety Board.

NTSB. 1994. US Air Flight 427: Aircraft Accident Report. Washington: National Transportation Safety Board.

NTSB. 2000. Aircraft Accident Report: Korean Airlines Flight 801. Washington DC: National Transportation Safety Board.

NTSB. 2001. NTSB Adopts Revised Final Report on the 1991 Crash of United Airlines Flight 585 in Colorado Springs, CO." In *United 585 Revised Report*. Washington, DC: National Transportation Safety Board.

NTSB. 2002. *Loss of Control and Impact with Pacific Ocean, Alaska Airlines Flight 261, McDonnell Douglas MD-83*. Washington, DC: National Transportation Safety Board.

NTSB. 2014. *Asiana Flight 214: Descent Below Visual Glidepath and Impact with Seawall*. Washington, DC: National Transportation Safety Board.

Pakistan CAA. 2011. *Investigation Report - Air Blue Flight ABQ202*. Karachi: Pakistan Civil Aviation Authority.

Reason, James. 1997. *Managing the Risks of Organisational Accidents*. Aldershot, UK: Ashgate Publishing Limited.

SAE. 2010. *Guidelines for the Development of Civil Aircraft and Systems ARP4754A*, edited by John Dalton. Society of Automotive Engineers.

Thaler, Richard H., and Cass R. Sunstein. 2008. *Nudge: Improving Decisions About Health, Wealth, and Happiness*. New York: Penguin Books.

Vaughn, Diane 1997. *The Challenger Launch Decision: Risky Technology, Culture, and Deviance at NASA*. Chicago, IL: University of Chicago Press. Original edition, 1996.

14 The Supply Chain

The supply chain is an area of high risk as will be explained in this section. This risk is increased by the consideration of second-tier suppliers.

WHAT IS THE SUPPLY CHAIN?

The supply chain is the sequence of suppliers who design and make subsystems and components for an aircraft. Each supplier is dependent on a procurement specification provided either by the OEM or a higher-level supplier. Figure 14.1 proves a schematic of that sequence. Chapter 6 explains procurement specifications and their importance.

The center circle is the OEM or the organization responsible for designing the entire aircraft and managing the suppliers. On either side is a first-tier suppler, that is, a supplier that reports directly to the OEM. Beyond each first-tier supplier is a second-tier supplier that reports directly to the first-tier supplier.

SUPPLY CHAIN RISKS

There are two primary supply chain risks. The first has to do with the length of the supply chain, and the other has to do with the management of second-tier suppliers. The supply chain is a major source of risk in the commercial aviation domain. Jackson (2015, pp. 178–180) provides a comprehensive explanation of supply chain risks and their treatment.

The length of a supply chain is a risk because the more suppliers there are means that there will be more organizational interfaces. Each organizational interface is a risk because this is where communications tend to falter.

There is a fallacy in many companies that longer supply chains will actually reduce risk. This is true to a certain extent. That is to say, financial risk may be lower, but technical risk may be substantially increased due to the increased number of organizational interfaces created.

Of course within risk analysis, technical risks may have financial consequences, so the premise that supply chains will reduce cost risk is highly suspected.

Second-tier risks. The very existence of second-tier suppliers creates risks. These risks occur because the OEM has no direct control over the procurement specification or statements of work (SOWs) that go to the second-tier suppliers.

The quality of second-tier products is dependent on two primary factors: the quality of the procurement specification and the quality of the statement of work. The flow-down of requirements to second-tier suppliers is always difficult.

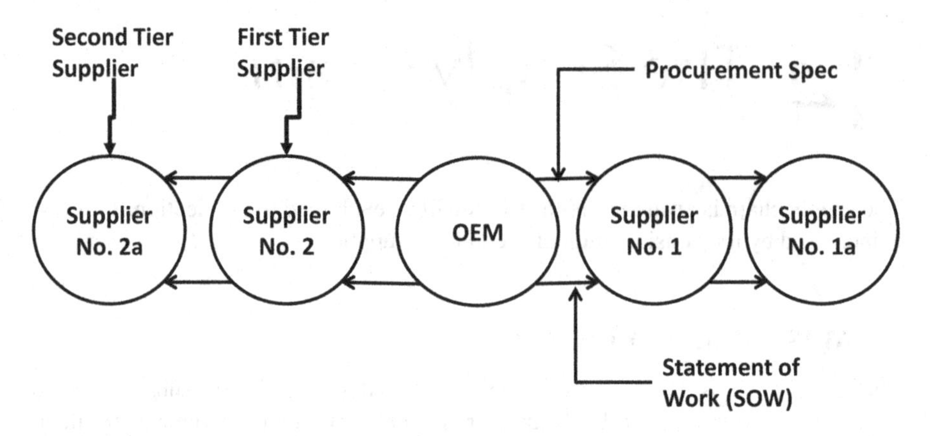

FIGURE 14.1 Supply chain sequence.

SUMMARY OF SUPPLY CHAIN RISKS

Of course quality products would always result if there were quality specifications and quality SOWs. However, the reality of the complexity of the supply chain makes this virtually impossible. Only quality management can achieve this goal.

REFERENCE

Jackson, Scott. 2015. *Systems Engineering for Commercial Aircraft: A Domain Specific Adaptation*, edited by Guy Loft, 201–213. Second ed. Aldershot, UK: Ashgate Publishing Limited (in English and Chinese). Textbook.

15 Aircraft Systems with Feedback Loops

One of the basic features of any system, according to Meadows (2008, pp. 25–34), is the ability of any system to run itself. It does this through the mechanism of information feedback loops. She says, "everything we do as individuals, as an industry, or as a society is done in the context of an information-feedback system."

For the purpose of illustration, using simplified diagrams, we will illustrate the principle of feedback through the pneumatic control system using the pitot tube to measure the aircraft's angle of attack and speed.

For the uninitiated a pitot tube is a slender tube on the front of an aircraft with two holes in it. One hole is to measure stagnation pressure, the pressure of the oncoming air flow. The other hole measures static pressure, the pressure exerted by the side flow. Together these measurements can measure the speed of the aircraft and the angle of attack.

One of the major problems with pitot tubes is the build-up of ice crystals as happened on Air France 447 off the coast of Brazil in 2009 resulting in the death of all occupants.

NOTES ON FIGURES AND SKETCHES OF FEEDBACK SYSTEMS

The figures and sketches in this section pertaining to feedback systems are notional depictions how feedback systems work and are not intended to be exact depictions of their operations. They are intended to be approximate depictions so that the reader can get a basic understanding of their principles of operation. The sketches in this section are intended to explain the principles behind the operation of an aircraft pitot tube. An actual pitot may differ in its details.

HOW A FEEDBACK SYSTEM WORKS?

As in the example of a pitot tube, the purpose of the device is to measure both how well the aircraft is performing and to make corrections whenever it is going too fast or too slow. Figure 15.1 illustrates both situations.

In the top part of Figure 15.1, the pitot tube sucks in air from the environment and measures the performance both in terms of speed and angle of attack. If the speed is too high or the angle of attack is too steep, it will adjust both to achieve the desired speed and angle.

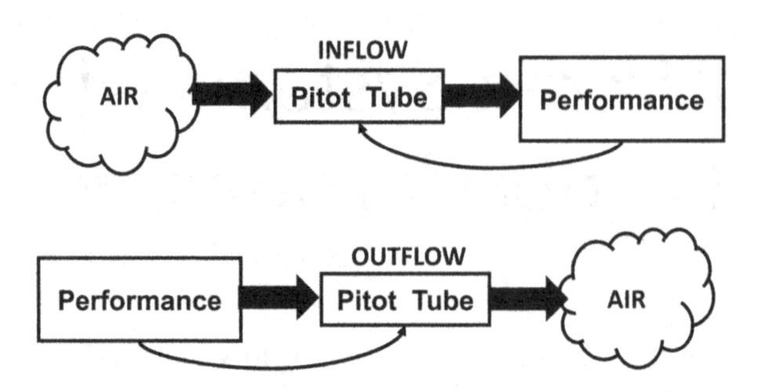

FIGURE 15.1 Pitot tube functioning.

In the lower part, the pitot will measure the performance outflow into the pitot tube. The objective of the pitot tube is to message the control system either to increase the speed or adjust the angle of attack until the desired speed and angle are achieved.

The performance of the pitot tube is highly sensitive to the flow in and out of the holes in the tube. If the holes are blocked, the tubes will misread the performance being achieved.

FEEDBACK LOOPS WITH BALANCING LOOPS

According to Meadows (2008, p. 30), "balancing feedback loops are equilibrating or goal-seeking structures in systems and are both sources of stability and resistance to change." In the example of Figure 15.1, the two pitot tubes operated independently trying to seek an optimum balance in performance and flight angle. In another example, balancing feedback loop, we call B, operates as a system, that is, it operates as a whole.

In Figure 15.2, the diagram shows the aircraft flying at a certain level of performance in both speed and flight angle. The pitot tube measures the current performance. The pitot tube "knows" that the aircraft is flying at a performance level less than its potential. The pitot can then measure that discrepancy between the actual and potential levels of performance. This discrepancy results in the balancing loop B. The pitot tube can then signal the aircraft to increase its performance level and reduce the discrepancy to zero. It is a single pitot tube that measures the discrepancy and signals the order for an improved performance.

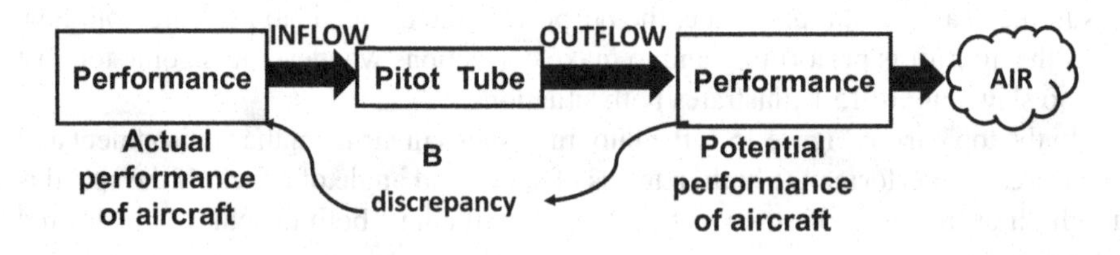

FIGURE 15.2 Balancing loops.

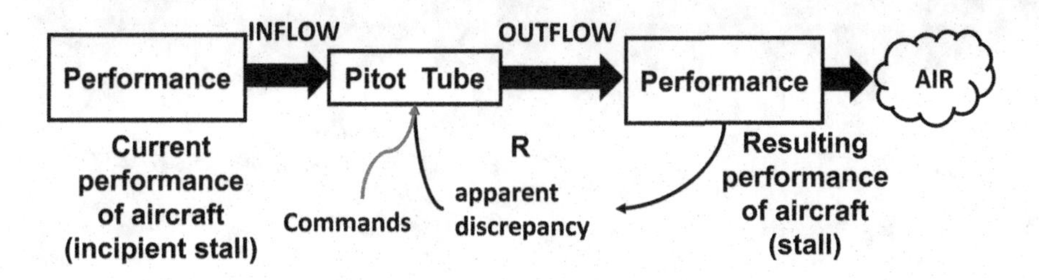

FIGURE 15.3 Reinforcing loops.

FEEDBACK LOOPS WITH REINFORCING LOOPS

Often feedback loops get out of control and sometimes lead to catastrophic consequences. This happens typically when the aircraft is on the verge of a stall, for example, at a high altitude and a high angle of attack. In this situation, increased performance will not always bring the aircraft back into stable flight.

Figure 15.3 illustrates this situation. In this diagram the aircraft is flying at "incipient stall," that is, at a high altitude and/or a high angle of attack. In this situation the pitot tube detects the need for increased performance (a flawed analysis). If the aircraft does initiate increased performance, it will initiate a true stall, losing lift, and losing altitude and control. The factors that cause this condition are the flawed commands given to the aircraft when it was in the incipient stall condition.

This situation, however, is correctible. An experienced pilot will know when the aircraft is in the incipient stall condition and can make the appropriate commands. Among these commands are actions to reduce the angle of attack and not gain speed. Increasing the angle of attack will almost always make the situation worse.

REFERENCE

Meadows, Donnella. 2008. *Thinking in Systems.* White River Junction, VT: Chelsea Green Publishing.

16 A Final Word

The simple reason that this book is written from a systems approach perspective rather than a systems engineering perspective is that the systems approach perspective is more comprehensive. It includes systems architecting, for example, which defines the arrangement of the parts of an aircraft, both physically and functionally, which systems engineering does not do. The systems approach defines the aircraft both from an architectural perspective and a performance perspective. In short, this book suggests that merging these two perspectives will result in superior aircraft.

In addition, this book examines a broader range of systems rather than just the aircraft itself. For example, it includes the air traffic control (ATC) system, the development system, the maintenance system, the aircraft subsystems, and the worldwide aviation system.

Next, this book discusses many aspects of systems identified by Bertalanffy (1968), Meadows (2008), and others. These include hierarchy, holism, emergence, feedback loops, and many other aspects.

Apart from the theoretical and philosophical aspects, also included are discussions of topics of interest to modern developer, such as cybersecurity protection and supply chain risk.

In short, this book has sought to provide the most comprehensive description of aircraft definition possible.

REFERENCES

Bertalanffy, Luwig von. 1968. *General Systems Theory: Foundation, Development, Applications*. Revised ed. New York: George Baziller.

Jackson, Scott. 2015. *Systems Engineering for Commercial Aircraft: A Domain Specific Adaptation*, edited by Guy Loft. Second ed. Aldershot, UK: Ashgate Publishing Limited (in English and Chinese). Textbook.

Meadows, Donnella H. 2008. *Thinking in Systems*. White River Junction Vermont: Chelsea Green Publishing.

Systems Approach Glossary with Aviation Examples

Topic	Chapter	Definitions	Example in an Aviation Context
Abstracted	1	A conceptual (or abstract) system that represents parts of the real world (Dori et al. 2019).	The aircraft reliability tree is an abstracted system.
Architecting	9	The arrangement of the parts of a system, either physical or functional.	The Blended Wing Body contains new approaches to both physical and functional architecting of an aircraft.
Assumption	5	An acknowledgment of the existence of a particular risk situation and a conscious decision to accept the associated level of risk without engaging in any special efforts to control it (Conrow 2003, p. 29).	Designers of aircraft will make the assumption that certain risks are too rare to design for, such as extreme levels of lightning.
Avoidance	5	Involves a change in the concept, requirements, specifications, and/or practices that reduce risk to an acceptable level (Conrow 2003, p. 30).	The use of nitrogen-enriched air in fuel tanks has reduced the risk of inflammation to an acceptable level.
Boundary	1	A distinction is made by an observer, which marks the difference between an entity and its environment (Checkland 1999, p. 312).	For commercial aircraft, the boundary of the system can be viewed as its farthest point of communication.
Cognitive bias	6	A systematic error in judgment and decision-making common to all human beings, which can be due to cognitive limitations, motivational factors, and/or adaptations to natural environments (Wilke and Mata 2012).	The *rankism* cognitive bias has been the root cause of several aircraft accidents.
Cohesion	1	The attribute of a system that allows it to operate before, during, and after an encounter with a threat (Hitchins 2009, pp. 59–63).	All the elements of an aircraft must function in cohesion to perform as expected.
Complexity (objective)	4	The extent to which future states of the system cannot be predicted with certainty and precision, regardless of our knowledge of current state and history (BKCASE Editorial Board 2016).	The complexity of the air traffic control system must be reckoned with.

(*Continued*)

Topic	Chapter	Definitions	Example in an Aviation Context
Constructivist	1	Systems are mental constructs that humans create in their minds to explain aspects of how the world works (Dori et al. 2019).	The flight manual is a pure constructivist system.
Control	5	Manages the risk in a manner that reduces the likelihood and/or consequence of its occurrence on the program (Conrow 2003, p. 30).	Flying only in daylight will control the risk of aircraft accidents.
Critical systems thinking	SEBoK	An approach to help select between hard system and soft system methods (Jackson 1985, pp. 135–151).	In the larger view, commercial aircraft will require both hard and soft system thinking.
Cybernetics	1	A subordinate discipline to systems science. The study and modeling of communication, regulation, and control in systems (Wiener 1948).	Cybernetics is essential to the understanding of the control of an aircraft.
Cybersecurity	10	Protection from external electronic adversaries.	Cybersecurity is an important consideration in modern aviation.
Emergence	1	The principle that whole entities exhibit properties only when attributed to the whole, and not to its parts (Checkland 1999, p. 314).	Flight is an emergent property of an aircraft.
Engineering	1	The action of working artfully to bring something about (Oxford 2019).	Within systems science engineering usually has this broader meaning.
Environment	1	That which lies outside the system boundary in the formal model (Checkland 1999, p. 314).	Continuous measurement of the weather environment is essential to safe flying.
Feedback loops	1	The transmission of information about the actual performance of a machine to an earlier stage in order to modify its operation (Checkland 1999, p. 55).	The aircraft control system contains feedback loops.
Function	1	The means, techniques, roles, behavior of an agent to the required task or subtask (Sheridan 2009, p. 650).	The function of the propulsion system is to provide thrust.
Hard systems thinking	1	To create a new system that can be introduced into some problematic situation to neutralize or solve the problem (Hitchins 2007, p. 23).	The design of the physical aircraft is the product of hard systems thinking.
Hierarchy	1	The principle to which entities meaningfully treated as wholes are built up of smaller entities which are themselves wholes. In a hierarchy emergent properties denote the levels (Checkland 1999, p. 314).	ATA chapters are a hierarchical view of an aircraft.

(Continued)

Topic	Chapter	Definitions	Example in an Aviation Context
Holism	1, 3	The theory that parts of a whole are in intimate interconnection, such that they cannot exist or be understood independently of the whole (Bertalanffy 1968).	An aircraft and its external connections must be considered holistically.
Human activity system	Web	A notional purposive system that expresses some purposeful human activity, which could in principle be found in the real world (Checkland 1999, p. 314).	Piloting an aircraft is a human activity system.
Individual bias	6	Biases suffered by individuals when they are making decisions (Jackson 2020).	Many aviation catastrophes can be traced to individual biases by the pilot.
Information theory	Web	The study of the information in a system (Checkland 1999, pp. 89–92).	Information theory provides insight into communications in aviation.
Interactions	1	Relationships that link elements of a system (Sage and Armstrong 2000, pp. 314–315).	The interaction between the pilot and the flight control system is a critical relationship.
Interdisciplinary	Web	People with a wide range of skills working together (Sillitto et al. 2018).	The traditional view of systems engineering is interdisciplinary.
Interface	4	A shared boundary between two functional units, defined by functional characteristics, common physical interconnection characteristics, or other characteristics, as appropriate (from ISO 2382-1) (INCOSE 2015, p. 263).	Aircraft have many interfaces both internal and external.
Irrationality	6	When beliefs are dominated by prior beliefs or emotion (Borelli 2016).	Irrationality is the root cause if many cognitive biases.
Mental systems	1	A system based on a person's perceptions (Sillitto 2014, p. 97).	A mental view of an aircraft.
Need		A lack of something that is desired or required (Armstrong 2009, p. 1035). A need is a prerequisite for defining system requirements.	An aircraft can be needed to transport passengers and cargo between sites.
Open systems	Singer	Defines elements and relationships which can be considered part of the system and describe how these interact across the boundary with related elements in the environment (Singer, Hybertson, and Adcock 2019).	Aircraft are open systems.
Operations research	1	A subordinate discipline to systems science. The use of mathematical modeling and statistical analysis to optimize decisions on the deployment of the resources under an organizations control (BKCASE Editorial Board 2016).	Operations research is used for air traffic management in the aviation domain.

(Continued)

Topic	Chapter	Definitions	Example in an Aviation Context
Organizational bias	6	Biases operative in an organizational context and result from specific characteristics of an organization (Jackson 2020).	A hierarchical organization structure is a characteristic that may result in the *rankism* bias.
Requirement		Description of how the system should behave, application domain information, constraints on the system's operation, or specification of a system property or attribute (Kotonya and Sommerville 1998, p. 6). System requirements are determined following the identification of system need.	The aircraft propulsion system shall provide 5,000 kg of thrust at an altitude of 20 km.
Resilience	7	The ability to maintain required capability in the face of adversity (BKCASE Editorial Board 2016).	The survival of 155 occupants is a demonstration of the resilience of US Airways Flight 1549.
Rich picture	9	The expression of a *problem situation* compiled by an investigator, often by examining elements of *structure*, elements of *process*, and the situation *climate* (Checkland 1999, p. 317).	A common problem situation for commercial is *extended range*; the structural elements may include *increased fuel capacity* and *wing length*; the process consists of conducting *fuel and wing length trade-offs*; the situation is *market demand*.
Risk	5	A measure of the potential inability to achieve overall program objectives within defined cost, schedule, and technical constraints (Conrow 2003, p. 21).	The *optimism* bias is a recognized risk on aircraft programs.
Safety	6	The expectation that a system does not, under defined conditions, lead to a state in which human life, health, property, or the environment is endangered (ISO/IEC NP/TR 33015 2014).	Safety is the primary concern of FAA regulations.
Soft systems thinking	1	To look for symptoms of dysfunctions in existing symptoms and seek to repair, or work around, the dysfunctions so as to suppress the symptoms (Hitchins 2007, p. 23).	Tracking the source of faulty aircraft maintenance requires soft systems thinking.
Supply chain	8	An integrated chain of systems in which an organization produces systems, products, and services to support another downstream system (Wasson 2006, p. 130).	An overextended supply chain can result in added risk.

(*Continued*)

Topic	Chapter	Definitions	Example in an Aviation Context
Synthesis	1	Searching for or hypothesizing a set of alternative courses of action or options (Sage and Armstrong 2000, p. 55).	A sound aircraft architecture requires a thorough synthesis process.
System	1	A traditional definition of system is "… *an integrated set of elements, subsystems and assemblies that accomplish a defined objective. These elements include products (hardware, software, firmware), processes, people, information, techniques, facilities, services, and other support elements*" (Dori and Sillitto 2017).	An aircraft is a human-made system.
System analysis	1	A subordinate discipline to systems science. A systematic investigation of a real or planned system to determine the information requirements and processes of the system, and how these relate to each other and to any other system (ISO/IEC/IEEE 2009).	System analysis can be used to determine the information required among nodes of the aviation system.
System dynamics		A subordinate discipline to systems science. A methodology based upon two primary loops (growth and limit) to describe a variety of situations (INCOSE 2015, p. 19).	Aircraft can operate in many situations that can be analyzed by system dynamics techniques.
System of interest	2	The system whose life cycle is under consideration (INCOSE 2015, p. 268).	For most engineers an aircraft is the system of interest.
Systems approach	1	A subordinate discipline to systems science. A means of identifying and understanding complex problems and opportunities, synthesizing possible alternatives, analyzing and selecting the best alternative, implementing and approving a solution, as well as deploying, using, and sustaining solutions (BKCASE Editorial Board 2016).	A systems approach is required to capture all aspects of an aircraft system.
Systems architecting	1, 4	The activity of creating a system architecture with the aim that the system will be built to do the job it was meant to do—in other words it will be "fit for purpose" (Sillitto 2014, p. 4).	The architecting of an aircraft system requires a broad knowledge of the aircraft's purpose.

(Continued)

Topic	Chapter	Definitions	Example in an Aviation Context
Systems engineering	1	A transdisciplinary approach and means based on systems principles and concepts to enable the realization of successful whole-system solutions (Sillitto et al. 2018).	The systems engineering of commercial aircraft requires the integration of many disciplines.
Systems of systems engineering	1	The engineering of systems that comprises other component systems, and where each of the component systems serves organizational and human purposes (Jamshidi 2009, p. 4).	Aircraft systems are systems of systems.
Systems thinking	1	A subordinate discipline to systems science. An epistemology that, when applied to human activity, is based on the four basic ideas: emergence, hierarchy, communication, and control (Checkland 1999, p. 318).	Systems thinking is the beginning point for viewing an aircraft and its external interfaces as a system.
Traceability (of requirements)	3	The linking of requirements to sources and components (Kotonya and Sommerville 1998, pp. 128–129).	Requirements of all components need to be traceable to stakeholder needs.
Transdisciplinary	2	Enables inputs and participation across technical and nontechnical stakeholder communities and facilitates a systemic way of addressing a challenge (Sillitto et al. 2018).	Transdisciplinary is the goal of future engineering of systems.
Transfer	5	May reallocate risk during the concept development and design processes from one part of the system to another, thereby reducing the overall system and/or lower-level risk or redistributing risks among stakeholders (Conrow 2003, p. 31) (adapted).	Risks are often transferred to suppliers with the associated financial aspects.
Vee model	3	A sequential method used to visualize various key areas for systems engineering focus, particularly during the concept and development stages (Sillitto 2014, p. 33).	The Vee model reflects the aspects of the aircraft development, including its hierarchy.
Wholeness, togetherness	Singer	The drawing together of various parts and the relationships they form in order to produce a new whole (Boardman and Sauser 2008).	An aircraft cannot function properly as a system unless all its parts exhibit wholeness.
Worldviews	1	A cognitive orientation of an individual toward a concept or set of concepts (Dori et al. 2019).	Surveys show that individuals exhibit different worldviews of systems, including aircraft.

REFERENCES

Armstrong, J. H. 2009. "Issue Formulation." In *Handbook of Systems Engineering and Management*, edited by Andrew Sage and William B. Rouse, 1027–1089. Hoboken, NJ: Wiley.

Bertalanffy, Luwig von. 1968. *General Systems Theory: Foundation, Development, Applications.* Revised ed. New York: George Baziller.

BKCASE Editorial Board. 2016. "Systems Engineering Body of Knowledge (SEBoK)." Accessed 1 May. http://sebokwiki.org/wiki/Guide_to_the_Systems_Engineering_Body_of_Knowledge_(SEBoK).

Boardman, J., and B. Sauser. 2008. *Systems Thinking—Coping with 21st Century Problems.* Boca Raton, FL: CRC Press.

Borelli, LIz. 2016. "The Psychology Behind Irrational Decisions Has a Lot to Do with How You Manage Emotions." *Medical Daily.*

Checkland, Peter. 1999. *Systems Thinking, Systems Practice.* New York, NY: John Wiley & Sons.

Conrow, Edmund H. 2003. *Effective Risk Management: Some Keys to Success.* Second ed. Reston, VA: American Institute of Aeronautics and Astronautics.

Dori, Dov, and Hillary G. Sillitto. 2017. "What is a System? An Ontological Framework." *Systems Engineering* 20(3):207–219.

Dori, Dov, Hillary G. Sillitto, Regina Griego, Dorothy McKinney, Eileen Arnold, Patrick Godfrey, James Martin, Scott Jackson, and Daniel Krob. 2019. "System Definition, System Worldviews, and Systemness Characteristics." *IEEE Systems Journal.* Preprint (accepted for publication).

Hitchins, Derek. 2009. "What are the General Principles Applicable to Systems?" *Insight* 12:59–63.

Hitchins, Derek K. 2007. "Systems Engineering: A 21st Century Systems Methodology." In *Wiley Series in Systems Engineering and Management*, edited by Andrew P. Sage. Hoboken, NJ: John Wiley & Sons.

INCOSE. 2015. *Systems Engineering Handbook*, edited by SE Handbook WG. Seattle, CA: International Council on Systems Engineering.

ISO/IEC NP/TR 33015. 2014. Information Technology: Process Assessment—Guide to Process Related Risk Determination. In International Standards Organization. Accessed 1 May. www.iso.org.

Jackson, M. 1985. "Social Systems Theory and Practice: The Need for a Critical Approach." *International Journal of General Systems* 10:135–151.

Jackson, Scott. 2020. *High-Risk Decisions: A Transdisciplinary Science.* Cockeysville, MD: TBD.

Jamshidi, Mo. (ed.). 2009. "Systems of Systems Engineering: Innovations for the 21st Century." In *System of Systems Engineering: Innovation for the 21st Century.* Hoboken, NJ: John Wiley & Sons.

Kotonya, Gerald, and Ian Sommerville. 1998. "Requirements Engineering." In *Worldwide Series in Computer Science*, edited by David Barron and Peter Wegner. New York, NY: Wiley.

Oxford. 2019. "Definition of Engineering." Oxford University Press, accessed 17 September. https://www.lexico.com/en/definition/engineering.

Sage, Andrew, and James E. Armstrong Jr. 2000. "Introduction to Systems Engineering." In *Wiley Series in Systems Engineering*, edited by Andrew P. Sage. Hoboken: John Wiley & Sons.

Sheridan, Thomas B. 2009. "Human Supervisory Control." In *Handbook of Systems Engineering Management*, edited by Andrew Sage and William B. Rouse. Hoboken, NJ: Wiley.

Sillitto, Hillary G. 2014. "Architecting Systems: Concepts, Principles, and Practice." In *Systems*, edited by Harold "Bud" Larson, Jon P. Wade, and Wolfgang Hofkirchner, Vol. 6. London: College Publications.

Sillitto, Hillary G., James Martin, Regina Griego, Dorothy McKinney, Dov Dori, Scott Jackson, Eileen Arnold, Patrick Godfrey, and Daniel Krob. 2018. "A fresh look at Systems Engineering – What Is It, How Should It Work?" IS 2018, Washington, DC.

Singer, Janet, Duane Hybertson, and Rick Adcock. 2019. "Introduction to System Fundamentals." In *SEBoK Extract*. San Diego, CA: INCOSE.

Wasson, Charles S. 2006. "System Analysis, Design, and Development." In *Wiley Series in Systems Engineering and Management*, edited by Andrew P. Sage. Hoboken, NJ: John Wiley & Sons.

Wiener, N. 1948. *Cybernetics: Or Control and Communication in the Animal and the Machine*. Cambridge, MA: The Massachusetts Institute of Technology Press.

Wilke, A., and R Mata. 2012. Cognitive Bias. In *The Encyclopedia of Human Behavior*, edited by V. S. Ramachandran. Cambridge, MA: Academic Press.

Index

Note: Boldface page numbers indicate primary reference.

V

W

Printed in the United States
by Baker & Taylor Publisher Services

Printed in the United States
by Baker & Taylor Publisher Services